A Culture of Recycling / Recycling Culture?

Literary and Cultural Theory

General Editor: Wojciech H. Kalaga

Vol. 37

PETER LANG

Frankfurt am Main · Berlin · Bern · Bruxelles · New York · Oxford · Wien

Wojciech Kalaga
Marzena Kubisz
Jacek Mydla
(eds.)

A Culture of Recycling / Recycling Culture?

PETER LANG
Internationaler Verlag der Wissenschaften
Frankfurt am Main · Berlin · Bern · Bruxelles · New York · Oxford · Wien

Bibliographic Information published by the Deutsche Nationalbibliothek
The Deutsche Nationalbibliothek lists this publication in the Deutsche Nationalbibliografie; detailed bibliographic data is available in the internet at http://dnb.d-nb.de.

Cover Design:
Olaf Glöckler, Atelier Platen, Friedberg

ISSN 1434-0313
ISBN 978-3-631-60163-1

www.peterlang.de

Contents

PART III. CHALLENGING REPETITION

PART IV. RECYCLING THEORY

Preface

The purpose of this volume is to address the notion of cultural recycling by assessing its applicability to various modes of cultural and theoretical discourse. The word "recycling" is here used, collectively to denote phenomena such as cyclicity, repetition, recurrence, renewal, reuse, reproduction, etc., etc., which seem to be inalienable from basic cultural processes.

From the Platonic ideals of the circle and circular movement, the idea of the wheels of Fortune and their ongoing applications in the artistic representations of the human condition, through that of the Nietzschean eternal recurrence and the Freudian return of the repressed right down to William B. Yeats' apocalyptic visions of the gyres – cyclicity and circularity seem to have predetermined the ways in which the Western mind represented itself. T. S. Eliot's conviction that "No poet, no artist of any art, has his complete meaning alone. His significance, his appreciation is the appreciation of his relation to the dead poets and artists," famously abolished the romantic belief in the self-standing, all-creating, auto-legislating prerogatives of the Poet. Walter Benjamin's conception of the "mechanical reproduction" and Harold Bloom's hypothesis of the anxiety of influence and the "return of the dead" have diagnosed a state of cultural exhaustion. The idea of recycling conceived as a cultural necessity administers what seems to be a lethal stab to conceptions of creativity prefigured at the dawn of modern times, as in Edward Young's praise of the Original Genius, opposed to mere imitators and propped up ideologically by dedication to the Novel at the birth of contemporary ideas of fiction, and legally, by copyright laws.

Is the idea of originality, for so long regarded as an author's lifeblood, as sole justification of literary production and Prime Mover of cultural circulation, dead and buried? Is contemporary culture fated to belabour What Once Was? Is it doomed to remain stuck upon a treadmill of recycling or caught in a vicious circle of repetition and reuse, if not crass reproduction? The contemporary anxiety over cultural exhaustion, simulation practices, plagiarism, and piracy makes us wish to revisit questions concerning the production and circulation of meaning in the conditions marked by, to borrow Salman Rushdie's phrase, "the bogey of authenticity." But then, are recurrence and repetition something to run away from? Should we shun, fear and despise the merry-go-round of cultural production? Should it cause disenchantment and frustration?

Part of our purpose in proposing this theme is a desire to trace, confront, interrogate, and theorise the surviving phantoms of newness and paradigms of creativity or dreams of originality and to consider the need, a necessity perhaps, to overcome or sustain them, and, further, to estimate the possibility of cultural survival if it turns out, as it may, that culture is forever to remain an endless recurrence of the same.

Editors

PART I

ETERNAL RETURNS

Leszek Drong

The Eternal Return of Veridical Rhetoric: Why Even Antifoundationalists Cannot Help Recycling Foundationalist Tropes

In lieu of introduction I would like to indulge in a reminiscence about a situation which will serve as an illustration of the main issue raised in the title of this essay. A few years ago I had an MA student who claimed that I was being inconsistent in my criticisms of her project. You probably know the type: clever, inquisitive, imaginative, and painstakingly persistent in her pursuit of the ultimate answers to the perennial questions of humankind. First she attended my lectures on the New Pragmatism in literary studies, where I pontificated about the absence of immanent meaning in the text, the volatility of our basic discursive categories, the inevitable historicity of the major certitudes that we live by, and suchlike. She dutifully imbibed my convoluted explanations of why essentialism cannot hold and what is wrong with formalist thinking. When I discussed Richard Rorty's neopragmatism and Stanley Fish's antifoundationalism, she nodded her head vigorously, evidently captivated by the intellectual charm and rigour of their theories. At every mention of a potential analogy between American literary theory and European developments connected with poststructuralism and postmodernism she was clearly anxious to produce a long list of sweeping generalisations, most of which had to do with the recent paradigm's skeptical attitude to rationality, logic and absolute truths. Then she enrolled in my MA seminar and started writing her MA project.

The problem with the project was that it was written in a style which is characteristic of the least intelligible passages in contemporary French philosophy. Do not get me wrong: the student was brilliant – her ideas often attained a level of abstraction comparable only with Derrida's or Levinas's writings – but there was no argument in her work, its structure was a nightmare, and clarity was the last item on her agenda. Of course, I took a lot of time to explain to her what was wrong with her writing, having frequent recourse to the basic rules of logic, the fundamentals of composition and appeals to common sense. I insisted that her project be perspicuous, better organised, carefully structured, and more accessible to the reader. At this juncture, to my astonishment, she levelled at me a torrent of accusations. Apparently, I was guilty of a major incoherence at the heart of my professional performance. Initially, I had encouraged my students to accept the commandments of the postmodern creed only to regress intellectually and try to rehabilitate the outmoded categories of truth, logic and clarity in the end. How could I preach antifoundationalism, historicism, and antiformalism in the classroom and then expect my students to abide by these already discredited categories during my office hours? How could I – approvingly – quote Nietzsche's relativisation of truth, Rorty's devastating critique of epistemology, and Derrida's skeptical attitude to reason as one of our central transcendental signifieds and then require logic, argument, and structural discipline from my students? Now, what follows is an attempt to answer those questions and parry the objections raised against what some would see as my serious professional misconduct.

My student's complaint may be redescribed as a theoretical problem. If a person holds a particular position on a given subject, is s/he obligated to act on her/his views in real-life situations? Is it possible to declare certain principles and then break them as soon as we engage in our everyday doings? Or, to put it in more formal terms, should not one's *praxis* be concordant with one's theory? Even if all these questions can be answered in the negative, there still remains one nagging issue, that of the reasons why we should choose to act in this rather that that way, why in workaday contexts we should, out of all the available courses of action, put a premium on being reasonable and commonsensical by appealing to acknowledged truths and criteria. Again to transcribe the issue into more theoretical language, the big question is why, when the time comes to act, even staunch antifoundationalists revert to a rhetoric which pays homage to truth, reason and foundations. The answer I eventually offer is exceedingly simple and yet its felicity depends on a number of interrelated arguments and illustrations which will focus on the distinction between the discourse of theory and the more practical articulations associated with interpretation of individual works or cultural phenomena. Since my discussion will be restricted to the intersection of philosophy, literary theory and linguistics specifically concerned with speech communication, I shall designate as theory the philosophical models, imperatives and definitions of basic positions available to the interpreter while the term practice shall refer to the critical activity of explicating, discussing and assessing particular phenomena. By the expression veridical rhetoric, mentioned in the title of my essay, I mean a use of tropes and argumentative structures which privileges the categories of truth, logic, and consistency, and holds them to be foundation stones in any discursive edifice. At first sight, the continued recycling of this kind of rhetoric in postmodern culture (if that is the word) may be surprising, but on closer inspection we can identify compelling reasons why in everyday contexts the culture is resistant to rarified philosophical talk about the relativity of all truths and the historical constructedness of all foundations.

On a practical (as opposed to philosophical or theoretical) level, speech communication depends on shared assumptions. What has happened in the human sciences over the last half century may be described as an assault on most of those assumptions. Under various banners, poststructuralists, postmodernists, antifoundationalists, historicists, constructionists and scores of other like-minded species have waged a war against the received notions of truth, certain knowledge, rationality, normalcy and objectivity. They have fulminated against the claims of science to possess access to objective and incontrovertible knowledge; they have questioned the possibility of addressing any issue from a disinterested perspective; they have stressed the significance of prejudice and belief in our perception of reality. Of course, none of this was invented in the second half of the twentieth century. Such big names as Darwin, Nietzsche, Freud and Wittgenstein had paved the way for the antifoundationalist turn in the humanities, and they, in turn, also drew on many sources and inspirations. Their writings contributed to a phenomemon which may be described as a massive backlash – setting in in the final decades of the nineteenth century – against the construal of truth as a master trope, elevated to that status by Plato and virtually worshiped since then.

Truth, if we strip it of its metaphysical pretensions, is a linguistic category. Traditionally, a discourse is said to be true when its statements correspond to what they are taken to represent. Our language reflects the world that we describe by means of words; on a very basic level of representation words are like icons which portray chunks of reality (although Charles Sanders Peirce would have described them as symbols). This view, often identified as representationalist, places a high premium on truth, rather than fiction, and prioritizes truth-claims as particularly valuable forms of expression. Our utterances acquire an air of authority and dignity if we have a record of producing true statements; telling lies is tantamount to being insane or malevolent. It takes tremendous courage to try to swim upstream by questioning the time-honoured value of truth the way Friedrich Nietzsche did in his writings. Nietzsche takes the bull by the horns and casts doubt on the correspondence between words and things – the central premise of representationalism. Let's refresh our memory of a passage which most poststructuralists, either wittingly or unwittingly, have incorporated into their way of thinking about language[1]:

> What is a word? The image of a nerve stimulus in sounds. But to infer from the nerve stimulus a cause outside us, that is already the result of a false and unjustified application of the principle of reason. ... The different languages, set side by side, show that what matters with words is never the truth, never an adequate expression; else there would not be so many languages.[2]

1 In this respect, Nietzsche's legacy is evident, although not always overtly acknowledged, in the writings of such different figures as Michel Foucault (a historian of ideas), Richard Rorty (a neopragmatist philosopher) and George Lakoff and Mark Johnson (linguists), who draw a parallel between truth and metaphor and claim that our account of truth should be free of "the myth of objectivism (according to which truth is always absolute truth)." George Lakoff and Mark Johnson, Metaphors We Live By (Chicago and London: The University of Chicago Press, 2003), p. 160.

2 Nietzsche, "On Truth and Lie in an Extra-Moral Sense" in The Portable Nietzsche (New York: Penguin Books, 1976), p. 45.

In another oft-quoted passage, Nietzsche proceeds to describe truth as "a mobile army of metaphors, metonyms, and anthropomorphisms" [3] thus erasing the distinction between fact-stating and rhetoric. Henceforth truth will be but a trope for our urge to establish a direct connection between what we say and what the world is really like. This urge in turn is motivated by our need to communicate, no matter whether or not the picture of reality we rely on in our verbal transactions is correct. After all, in communication what matters in the first place is that it be effective.

It might seem that those who have embraced Nietzsche's skepticism about truth and about the veridical rhetoric which assumes truth's supremacy in public debate will translate this skepticism into their discursive practice. Especially in the field of interpretation, it is tempting to look for traces of epistemological relativism in the writings of such card-carrying antifoundationalists as Richard Rorty and Stanley Fish. However, when one examines the rules that they rely on in drawing their interpretive conclusions, it is evident that the thrust of their arguments is never diluted by doubt about the validity of their own truth claims. As an illustration, I would like to discuss Stanley Fish's position on the theory-practice nexus and then enlarge on his self-identification as a weak antifoundationalist.[4] Ultimately, his essays will also furnish an important clue to the question about the reasons why we cannot help recycling veridical rhetoric even when we profess antifoundationalism as a general view.

In a recent version of his spectacular "no-consequences" argument,[5] Fish raises the issue of philosophy's relationship with practice and concludes that it is virtually non-existent. Under a very telling title "Truth but No Consequences: Why Philosophy Doesn't Matter," he argues against making transitions from one's general philosophical position to the imperatives that one acts upon in mundane situations, including one's interpretive practice. In a rhetoric reminiscent of the later Wittgenstein, Fish describes theory and practice as two different language games. Philosophy is thus denied its superior status as a prescriptive discourse and he claims that the only way in which it can be related to practice is by feeding on it for empirical data. Therefore, Fish concludes, antifoundationalism does not commit one to any form of practice and does not preclude anyone from assuming

3 Nietzsche, "On Truth and Lie," p. 46.

4 See Gary A. Olson and Lynn Worsham, eds., Postmodern Sophistry: Stanley Fish and the Critical Enterprise (Albany: State University of New York Press, 2004), p. 281.

5 Fish's argument about the inconsequential status of theory (that is, its irrelevance to our interpretative practice) appears in his book Doing What Comes Naturally: Change, Rhetoric, and the Practice of Theory in Literary and Legal Studies (Oxford: Clarendon Press, 1989), particularly pp. 315-467. I discuss this argument at more length in Disciplining the New Pragmatism: Theory, Rhetoric, and the Ends of Literary Study (Frankfurt am Main, Berlin, Bern, New York, Paris, Wien: Peter Lang, 2007), pp. 129-135.

foundations in everyday situations.[6] To make this point clearer, he uses his own case as an example:

> I might have a radically textual view of literary interpretation and believe that the establishment of literary meaning is at bottom a matter of rhetoric, but when I open my copy of *Paradise Lost* and begin to read it, meanings – definite, clear, perspicuous, undoubted meanings – just leap out of the page at me; and if someone who saw other meanings thought to argue against me by reminding me of my general belief about the bottom-line rhetoricity of literary readings, I would dismiss his argument as being beside the point because the meanings I saw would have been the product, not of my general belief, but of the disciplinary and institutional investments [connected with the field of literary studies].[7]

In other words, what Fish draws on while interpreting Milton is a set of imperatives and rules of thumb recognized and sanctioned by other literary critics as good tools to "do interpretation." The grand theoretical vocabulary used to describe the scope of interpretation, its constraints and ontological status, is necessarily set aside when a different language game (that of interpreting a particular text, rather than discoursing on general hermeneutics[8]) is in progress. Despite his claim that no consequences follow from the theoretical/philosophical views we hold, Fish does offer some consolation to those who are at sea when they reach this stage in his argument. Namely, he says that just because theory provides no methodological aid to practitioners of interpretation, its antifoundationalist version can do them no harm, either. The claim that the universe is not governed by any rational principle which could be translated into the language of science, and that there are no ultimate certainties in which to ground our inquiry, does not paralyse our everyday performance. As Fish puts it, "your awareness, even knowledge, that the routines you are running and the evidentiary procedures you rely on and believe in are features of a contingent and revisable practice, of a practice that is, as they say, 'socially constructed,' will in no way erode the confidence with which you run those routines or generate that evidence."[9] It is possible to run those routines with confidence precisely because no metacritical account of our practice can affect the way we perform within a real-life context. Thus Fish reads a positive message into

---

6 See Stanley Fish, "Truth but No Consequences: Why Philosophy Doesn't Matter," Critical Inquiry 29 (Spring 2003), p. 412.
7 Fish, "Truth but No Consequences," p. 410.
8 Elsewhere Fish describes it as "the theory game" and claims that it is "fun to play and can even yield winners and losers, but its relevance to the world of practice is nil unless it is so highly mediated that it is no longer theory at all." Stanley Fish, "Theory Minimalism," San Diego Law Review 34 (2000), p. 764.
9 Fish, "Theory Minimalism," p. 767.

his conclusion that the discourse of theory has no claim on us outside the seminar room of the philosophy department.

If there is no link between our general views and our practical doings, it does not make any sense to talk about their consistency or inconsistency. As different language games, they do not have to be compatible and their agreement is a matter of pure coincidence. Fish insists that apparently strange and unexpected combinations can occur within a single person; there is nothing surprising about performing in a markedly different way in different contexts.[10] Again, he resorts to his own career for illustration: "You can give all the standard answers to all the pragmatist questions and still be an authoritarian in the classroom, a decided conservative in cultural matters or inclined to the absolutes in theology. (I am, in differing degrees, all three.)"[11] Likewise, Fish's argument, in the form of an appeal to his own personal experience, does not seek to anchor his assumptions to a philosophical doctrine or a general theory of the self. It proceeds from the empirical particulars of practice to a larger claim about other people and the way they act. Unlike theory, his reasoning is descriptive rather than prescriptive, which is why it is no paradox that his personality involves all of the above.

Now, Fish's formidable personality aside, let us return to truth and its role in contemporary antifoundationalist writings. In Rorty's works, truth makes a frequent appearance in the context of relativism, which Rorty is often accused of. In the introduction to his recent book *Philosophy and Social Hope*, Rorty notes that relativism has been particularly closely associated with the pragmatist redescription of truth – either, along Nietzschean lines, as "a mobile army of metaphors and metonyms," or, in accordance with William James's view, as "the expedient in the way of thinking."[12] These redescriptions of truth are taken to render it relative by leaving it at the mercy of wayward interpreters or subordinating it to the provisional drift of our convictions and assumptions. "The pragmatist relativisation of truth" is obviously a possible take on what resulted from James's redefinition of this notion but we must bear in mind that James did not really want to relativise truth; he wanted to replace its earlier essentialist understandings with his own. In

10 This argument has much in common with Richard Rorty's notion of the private/public split which characterizes the figure of the liberal ironist introduced in his Contingency, Irony, Solidarity (Cambridge: Cambridge University Press, 1989). A detailed discussion of the split can be found in Leszek Drong, "The Irony and Contingency of Richard Rorty's Ironism," in Ryszard Wolny, ed., Culture, Literature, Language. Wszechnica Świętokrzyska Periodical Publications 23/1 (Kielce: Wszechnica Świętokrzyska, 2006), pp. 53-66.

11 Stanley Fish, "Truth and Toilets: Pragmatism and the Practices of Life," in Morris Dickstein, ed., The Revival of Pragmatism: New Essays on Social Thought, Law, and Culture (Durham and London: Duke University Press, 1998), p. 426.

12 See Richard Rorty, Philosophy and Social Hope (London: Penguin Books, 1999), p. xiv. As a matter of fact, in the passage in which James defines truth along these lines he does not use the noun but its nominalised adjective (the nominal) – "the true." See William James, Pragmatism (New York: Dover Publications, 1995), p. 86. Interestingly, in his introduction to Philosophy and Social Hope, Rorty misquotes James's definition and replaces "thinking" with "believing."

consequence, his definition of truth appears relativist only in the light of how truth was construed until the beginning of the 20th century. If we adopt James's alternative construal of truth, the (once scandalous) relativism inherent in his position will no longer appear to be a failure but a simple reflection of his reliance on context and relations (hence "relativism") to other key categories in the pragmatist's vocabulary. This is exactly the strategy that Rorty follows in *Truth, Politics and 'Post-Modernism'*: he proposes a new understanding of truth, an understanding embedded in the context of his pragmatist views at large and yoked to the particular political programme which he is proffering. Rorty deflects the charge of relativism by pointing out that postmodernists (in this context sharing the pragmatist view of truth) believe, just like representationalists, that certain statements are true while others are not; they also "recognize that some beliefs are more reliable tools than others, and that agreement on which tools to use is essential for social cooperation."[13] Here, then, is how Rorty describes the bone of contention between representationalists and postmodernists:

> The real difference between the two camps is not between believing in Truth and disbelieving in it but between regarding the word "Truth" as the name of something which deserves to be loved and regarding it as an unhappy hypostatized adjective, a misleading reification. For the so-called "post-modernists," the adjective "true" is a perfectly useful tool, but the use of the noun "Truth" as the name of an object of desire is a relic of an earlier time: the time in which we believed that there was a natural order to be grasped.[14]

> They [postmodernists] say instead that certain metaphors which we once used to explicate the notion of truth – those which revolve around notions like correspondence and adequate representation – need to be abandoned. Doing so will lead us to stop playing a certain language-game: the language game which uses the hypostatized adjective 'truth' in such phrases as "the quest for Truth" or "the love of Truth."[15]

In fact, what Rorty does in the two passages quoted above is *relativise the representationalist/essentialist approach to truth* by demonstrating that this approach is not founded on unquestionable and unimpeachable rules and universals. It is one of several different notions of truth[16] – a point which by itself demolishes the basic principles of this approach. Furthermore, Rorty's strategy of deflecting the charge of relativism involves a reconstrual of the love of Truth as "simply intellectual curiosity, a virtue which goes along with tolerance and lack of

---

13 Richard Rorty, Truth, Politics and 'Post-Modernism' (Assen: Van Gorcum, 1997), p. 21.
14 Rorty, Truth, Politics and 'Post-Modernism', p. 21.
15 Rorty, Truth, Politics and 'Post-Modernism', p. 22.
16 For a detailed discussion of a whole range of various philosophical theories of truth, see Lawrence E. Johnson, Focusing on Truth (London and New York: Routledge, 1992).

fanaticism,"[17] a virtue which he also identifies with conversability – another major item in his final vocabulary. Moreover, in "Pragmatism, Relativism, Irrationalism" included in *Consequences of Pragmatism*, Rorty proffers the most straightforward repudiation of relativism in politics and real life. In his categorical refutation of relativism as "the view that every belief on a certain topic, or perhaps about *any* topic, is as good as every other,"[18] we detect the seeds of the private/public split introduced in *Contingency, Irony, and Solidarity*. Just like Fish, Rorty argues that it is necessary to differentiate between one's attitude towards *philosophical* theories and *real* ones.[19] Consequently, one can possibly be a metaphysical relativist but when it comes to specific real-life issues relativism simply does not enter the picture: "Nobody really cares if there are incompatible alternative formulations of a categorical imperative, or incompatible sets of categories of the pure understanding. We *do* care about alternative, concrete, detailed cosmologies, or alternative concrete, detailed proposals for political change."[20] Rorty's refutation of relativism based on the distinction between philosophy and real-life situations testifies to his commonsensical approach to veridical rhetoric, which he never disavows in his own discourse.

That is probably because, regardless of its theoretical moorings, truth has acquired a massive capital of positive connotations. Almost invariably, a discourse which unfurls veridical banners – including verification, empirical evidence, correspondence to how things actually are, sincerity, irrefutability of its claims (possibly grounded in some sort of transcendence), appeals to common sense, etc. – is likely to garner support and acceptance of the audience. In contrast, a speech, or a text, which flaunts doubts about its own status, and is marked by excessive reservations and lack of confidence in its own claims, does not stand much chance of being taken seriously and appreciated. In their work concerned with metaphor, Lakoff and Johnson note that in public life truth is aligned with power, and thus to relinquish the former is to deprive oneself of the latter. More than that, the discourse of truth often ensures our well-being by offering us a ticket to a harmonious existence within our community. Here is how Lakoff and Johnson explain truth's importance in our daily lives: "We base our actions, both physical and social, on what we take to be true. On the whole, truth matters to us because it has survival value and allows us to function in our world."[21] The veridical rhetoric which we use in mundane contexts is simply the most effective way of attaining our goals, including persuading others to our point of view, effecting changes in the way things are (or preserving the status quo, as the case might be), giving us an

---

17 Rorty, Truth, Politics and 'Post-Modernism', p. 23.
18 Richard Rorty, Consequences of Pragmatism (Minneapolis: University of Minnesota Press, 1982), p. 166.
19 See Rorty, Consequences of Pragmatism, p. 167.
20 Rorty, Consequences of Pragmatism, p. 168.
21 Lakoff and Johnson, Metaphors We Live By, p. 160.

opportunity to seize more power and the like. At the same time, veridical tropes are a password to public credibility and respectability. Of course, in most cases we judge people not only by their words but also by their deeds, yet much depends on how we are affected by their discourse. Still, I would be shooting myself in the foot if I claimed now that the premium we put on truth in speech communication is due to its fundamental ontological or epistemological advantages over other categories. Truth-related tropes and figures are not inherently superior to any other metaphors we live by; their major advantage lies in their common recognition, in their widespread occurrence and a prestige sanctioned by custom and tradition (that is, by accidents of history), and that is what gives them a special status among other forms of expression. Interpersonal communication is, of necessity, consensual. We can understand each other only because we have accepted a compromise on the rules of communication, which makes us assume shared meanings, recognize typical contexts and ascribe shared motives to every utterance. The assumption of a bipolar criterion – true vs. untrue – seems to be fundamental to how we interact verbally with our environment, no matter whether we believe that the world is flat, round or merely an illusion. At the end of the day, to communicate our philosophical findings, we will invariably resort to veridical rhetoric for the simple reason that we believe our conclusions to be true and we want others to share this belief with us.

Katarzyna Nowak[1]

The Myth of Eternal Return: Melancholic Formation of Identity and Production of Cultural Icons

How does one start to write about the ghostly presence of protagonists who are unreal in at least a double sense? How does one conjure up their presence so that their elusive existence is solid enough to be captured in a text? I am thinking about those characters in novels that were summoned to appear in a text as ghosts of their real life equivalents, from the very beginning not meant to imitate life but rather to imitate fiction. Two examples emerge: Maryna Zalenska and Norma Jeane, the former the protagonist of Susan Sontag's epic novel *In America,* which is an account of the Californian endeavor of a Polish actress and icon of the stage, Helena Modrzejewska; the other a character in Joyce Carol Oates's novel *Blonde*, which renders in a similarly fictionalized form the life and death of Hollywood icon Marilyn Monroe.

The two literary examples will serve to present different approaches to exile, and the differentiation begins in the very way the authors of the books reveal their methods of work with the material that is then molded into fiction. In a note preceding *In America,* signed S.S., we read:

> The story of In America is inspired by the emigration to America in 1876 of Helena Modrzejewska. ... Inspired by … no less and no more. Most of the characters in the novel are invented, and those who are not depart in radical ways from their real-life models.[2]

The author of this note did not bother to conceal her identity; she performs the authorial role, voicing her independence and creativity. The emphasis is on inspiration, and what is also stressed is the question of departure: obviously the author meant to point out the difference between a biographical and fictional account, yet what gets revealed, somehow uncannily, is the issue of deferral, the spatial differentiation, which is marked in the phrase describing the departure of the characters in the novel from their real-life counterparts.

What a reader finds in J. C. Oates's book is a presentation of similar concerns, though with the emphasis on different aspects:

> *Blonde* is a work of fiction. While many of the characters portrayed here have some counterparts in the life and times of Marilyn Monroe, the characterizations and incidents presented are totally the products of

1 The author is a grantee of the Foundation for Polish Science.
2 Susan Sontag, In America (New York: Picador, 2000).

> the author's imagination. Accordingly, *Blonde* should be read solely as a work of fiction, not as a biography of Marilyn Monroe.[3]

We read the above in a note preceding the novel itself, on the page with the publisher's note, unsigned. Here, the stress is on the reader, with a clear instruction of how the novel should be read, and what is the "correct" approach to the material presented. The fact that the note remains unsigned, combined with the place where it is printed, adds to the sense of its being an absolute, and objective, command. What the author recommends here is to observe the central protagonist of the novel as doubly removed from the real life: Norma Jeane is a character who is not meant to be the fictional representation of Marilyn Monroe; she is a character who is supposed to be a fiction of a fiction.

The question of deferral appears crucial in this context, combined with the melancholic formation of identity, where the two create a space which is unreal yet reveals the fissures in reality, or, in other words, in the ways we conceive of the world around and inside us. This space I will understand as one connected with the turn towards the maternal, yet paradoxically, one that can be reached only with the separation from the maternal body, which by Lacan is associated with the entrance into language.[4] The space created for the characters in those two novels will be the testing ground for the very idea of exile, for the possibility of departure from the dominant economy of language and identity.

It comes as no surprise, of course, that America will be designated as the space of the immigrant dream, the space where one can escape all previous limitations. Yet the break from the past is not always imagined as desirable, as we are reminded in Franz Kafka's *America.* In the opening passage the Statue of Liberty is described as holding a sword instead of a torch. Not a benevolent but rather a menacing symbol of the place, it can be read as representing a threatening, castrating Father, suggesting that there is no escape from the masculine topography and economy of the language.

However, if one still dreams of escaping linearity and phallogocentric order, America seems to play the role of the new space perfectly. How is this space characterized in the two novels? The points of correspondence seem curiously similar, with the emphasis on the sense of a break with the past, yet an unreal one. It is the space where one is removed from the past, or rather, creates a new sense of beginning and thus attempts to escape the linearity of time. The levels of deferral pile up, because it is not enough to go to the United States as such, one has to go further, and in both discussed novels this place is California. In Sontag's novel, the depiction is quite straightforward: as one of the characters wonders, "Doesn't it

---

3 Joyce Carol Oates, Blonde (New York: Ecco, 2000).

4 Jacques Lacan, "The Mirror Stage as formative of the function of the I as revealed in psychoanalytic experience" (1949), in: Ecrits: A Selection (New York & London: W.W. Norton and Company, 1977), p. 2.

seem very American ... that America has its America, its better destination where everyone dreams of going?,"[5] not only commenting on the space itself, but also illustrating the character of the immigrant nation, in which everyone envisages a space of ending, of a final destination. Even more explicitly, it is described in the following words: "California, the ending, the last beginning."[6] The emphasis is on the dream quality of the place, where anything is possible, and the American dream can be carried out, pushed to the limit, though it never in fact comes true. Described as such, it becomes the boundary of one's vision, which reminds one of death, which can also be imagined as the space and time of abolition of the categories of possibility and limit.

As is usually the case with writing about death, one borders on the naïve and the banal, which testifies all the more to the impossibility of conceptualizing one's mortality. It is only possible to write of death on the terms of life. It is the negative description of what death is not: it cannot be contained in words, because language remains the proof of life, even if we accept its being borne out of the death drive. So it seems that the movement represented when it comes to writing about death is circular, that is, it means going round in circles back to life, and curiously enough, for the protagonists of *In America,* it seems that the land that stands for what is customary and recognizable, that is, that represents life, is Poland: "Poland was circles – everything familiar, saturated, centrifugal," whereas America possesses the qualities of an unknown territory: "Here the country, ever more spacious and thinly marked, streamed and spiked in all directions."[7] The latter is a country in which the very possibility of death is denied; yet at the same time it presents death as an option, as the only option of survival, so it seems.

In the case of Poland circularity means familiarity, whereas in the case of California it means just the opposite, which is revealed in the episode when Zalenska discusses a play in which she is to act the role of Frou-Frou, who dies at the end. Peter, Zalenska's son, wants a different ending to the story: "Why does Frou-Frou have to die? ... She could jump up and say, I changed my mind. ... Then she could go out to California and go up in an air-ship and say, Try to catch me if you can."[8] Here the naïve character – which reminds one of the Romantic Wordsworthian notion of "the child is father of the man" – is telling the truth about the meaning of California, the place where one can escape even the limitations of death.

Yet this ultimate place seems to be the Protestant vision of heaven: it is described as "the Laborer's Paradise."[9] Being at the same time the space of a dream factory, Hollywood, it is supposed to unite two conflicting tendencies. Endowed

5 Sontag, In America, p. 120.
6 Sontag, In America, p. 327.
7 Sontag, In America, p. 313.
8 Sontag, In America, p. 328.
9 Sontag, In America, p. 121.

with the quality of a paradox, it has to be read in a circular fashion, in which the answer points towards the question. How is California characterized? "Salubrious Climate. Fertile soil. / No severe winters. No lost time."[10] If the time cannot be lost, then it cannot be "had"; it has to cease to exist altogether. California, then, must be the land of eternity, and hence the land of death.

Such a vision of the place agrees with its presentation in *Blonde*. The main protagonist dies at the end, and it is not only welcome event, but also anticipated from the very beginning, when we read in the opening passage: "There came Death undeterred by the smoggy spent air of Los Angeles. By the warm radioactive air of southern California where Death had been born."[11] The circular motion: going back to the event forestalled suggests not only the obvious inescapability of death, but being stuck on it and reviving the melancholic longing for the lost object.

I see the loss as connected with artistic performance as well as with the entrance into the language in Lacanian sense. In order to be able to use the language, which is the tool of creativity for the figures of the actresses discussed, the person must be separated from the mother. The condition for entering the language is the rejection from the source of primeval sense of oneness, and this traumatic process reminds one of the hardship of immigration. The loss of one's linguistic environment is thus characterized by Kristeva: "The foreigner, thus, has lost his mother."[12] The fact that it is not both parents that one is supposed to lose in the process, but the mother, is fortified by the phrase "mother tongue," in which emigration and initiation into the language are combined. In Polish, however, the corresponding phrase will be "father tongue," which seems to fit better the situation of the immigrants in Sontag's novel, who leave Poland and their Polish "father tongue." For the two discussed characters, the loss of paternal support and entrance into a foreign language start from different points of departure.

Kristeva illustrates the predicament of being a foreigner as such: "Not speaking one's mother tongue. ... Thus, between two languages, your realm is silence,"[13] and she adds, "Silence has not only been forced upon you, it is within you: a refusal to speak ... . Nothing to say, nothingness, no one on the horizon."[14] This loneliness, being devoid of the ability to use the language, stretches to infinity: the horizon is empty, the world silent.

In the narration in Sontag's book, the main protagonist remains mysteriously devoid of voice in the last scene. The whole chapter gives voice to Edwin Booth, an actor who plays with Zalenska, while she is given one line only to show mercy to the mean, drunk, rambling Booth. In a monologue he describes his life and acting career, among other issues. The reader knows the reaction of

---

10 Sontag, In America, p. 121.
11 Oates, Blonde, p. 3.
12 Julia Kristeva, Strangers to Ourselves (New York: Columbia University Press), p. 267.
13 Kristeva, Strangers to Ourselves, p. 275.
14 Kristeva, Strangers to Ourselves, p. 276.

Zalenska, but not what she says. And it is Booth who tells her she speaks with the accent she has been trying so hard to eliminate. In this gesture she is relegated to the place of a foreigner and is meant to remain there. Nothing can change the fact that she has lost her mother – "father" tongue. Even though she immigrated to California, the land of fantasy, this fantasy is bound to remain unfulfilled.

The common theme for the two novels is designating California as a place associated with death, separation, and, ultimately, with freedom. The question of artistic freedom is obviously crucial in the case of the two protagonists, both of them actresses. In order to achieve this freedom, they have to abandon the space governed by the paternal law, the "Name of the Father," and set out for the place where they can create their identities without the parental support. The loss of the mother, and consequently, the loss of the mother tongue remain central to the performance of their identities. Yet they remain trapped in the play of imagined loss and grief, hence one can designate their identity as melancholic. The two figures present two ways of coping with the loss. According to Julia Kristeva, one of the two modalities that permit the survival of rejection is oralization, which is associated with "melody, harmony, rhythm"[15] and characterized by "a reunion with the mother's body, which is no longer viewed as an (...) expelling and rejecting body, but rather as a vocalic one."[16] Maryna Zaleska serves as an example of this modality, whereas Norma Jeane will be presented as opting for the second modality, thus described by Kristeva: "The second modality ... appears in the reunion with brothers' bodies, in the reconstruction of a homosexual phrathry."[17] It is worth mentioning that one modality is "always inseparable"[18] from the other, thus both create a vision of coping with being rejected by the mother's body and entrance into language.

In Zalenska's case, the impossibility of speaking without an accent, impossibility of entering the language fully, and being reminded of one's alien status, testifies to the way she uses the language in terms of oralization. Language is central to her performance as a foreigner. She longs to belong, to see the body of the mother as vocalic, yet she can never succeed. Monroe remains an outsider, too. What is emphasized in her case is her incessant repetition of shots over and over again: here circularity serves as the sign of not belonging. She is not a foreigner in the strict sense of the word, yet her attempt to deal with the rejection of the mother's body is as unsuccessful. Death is presented in both cases as the way out of the circular motion dictated by the will to deal with the separation.

The silence imposed on both characters is strengthened by the circularity of their emergence as cultural icons: the real life characters are used as material for the characters of biographies, then into personas in fiction, and finally, are

15 Julia Kristeva, Revolution in Poetic Language (New York: Columbia Press, 1984), p. 80.
16 Kristeva, Revolution in Poetic Language, p. 79.
17 Kristeva, Revolution in Poetic Language, p. 79.
18 Kristeva, Revolution in Poetic Language, p. 79.

combined with their existence as icons of stage or screen. The effect of silence, their removal from life, the deferral, is finite.

Quite paradoxically, then, the price of an attempted escape from the Law of the Father and the symbolic order is the loss of mother/tongue, which is, simultaneously, the very condition for freedom, which proves to be impossible. Being stuck on this loss produces melancholic desire for oneness, which means repeating the gestures that are supposed to free us from the painful feeling of a loss. Desire for oneness means desiring death. And this death is realized in California.

Grzegorz Moroz

From a Theodrome to the Dance of Shiva-Nataraya – Recycling Aldous Huxley's Views on Circularity in Nature and Culture

I would like to begin re-cycling Aldous Huxley's views on circularity in nature and culture with two contrastive images from two of his novels, one written at the beginning and the other at the end of his career as a novelist. *Antic Hay* is his second novel and was published in 1923, and *Island* is his final novel, published a year before his death in 1962. *Antic Hay* opens in this way:

> Gumbril, Theodore Gumbril Junior, B.A. Oxon, sat in his oaken stall on the north side of the School Chapel and wondered, as he listened through the uneasy silence of half a thousand schoolboys to the First Lesson, pondered, as he looked up at the vast window opposite, all blue and jaundiced and bloody with the nineteenth century glass, speculated in his rapid and rumbling way about the existence and the nature of God.[1]

Gumbril is the second, after Denis Stone from *Crome Yellow*, in the long succession of Huxley's characters portrayed in a usually more than less denigerating manner bearing more or less direct similarities with Huxley himself. Gumbril is a disheartened and disgruntled public school teacher, as Huxley was at Eton from September 1918 to February 1919. But what is more important, Gumbril is a sceptic and an agnostic. So, when Reverend Pelvey booms that "the Lord our God is one Lord," we encounter this flippant, theologically philological free indirect speech:

> Our Lord; Mr Pelvey knew, he had studied theology. But if theology and theosophy, then why not theography and theometry, why not theognomy, theotrophy, theotomy, theogamy? Why not theo-physics and theo-chemistry? Why not the ingenious toy, *the theotrope or wheel of gods?* Why not *a monumental theodrome*?[2]

Gumbril's long, *in service* speculations about "the existence and the nature of God," which meander beyond good and evil or „merely below them," through the painful memories of the death of his mother who was good, "not nice, not merely *molto simpatico* ... but good," end with the conclusion: "But this was nonsense, all nonsense. One must think of something better than this."[3] And so he does; the

1 Aldous Huxley, Antic Hay (New York: Harper & Row, 1965), p. 3.
2 Huxley, Antic Hay, p. 3; my italics.
3 Huxley, Antic Hay, pp. 4-6.

hardness of oaken benches, combined with the length of Reverend Pelvey's service and Gumbril's own inborn "boniness" bring about the realization that "The real remedy, it suddenly flashed across his mind, would be trousers with pneumatic seats. For all occasions; not merely for church going."[4]

And on the strength of the hope in the commercial success for what becomes known as "Gumbril's Patent Small Clothes," Gumbril Junior quits his history teaching job and goes to London. He fails to make any money on his trousers, which is by no means the only failure in the novel, a novel about which Keith May observed:

> The shape of Antic Hay suggests a diagram of futility, a wheel at whose centre the hero cannot stay. Gumbril begins on the circumference, and the wheel is moving very slowly, a treadmill.[5]

The futility of modern, post-Great War life, its *waste land* aspect is high-lighted again in the final chapter of the novel, with the recurring motif of Shearwater, a scientist pedalling on his stationary bicycle. Today, when pedalling on a stationary bike, at home or in some fitness or wellness centre, is generally considered to be one of the best cardio-mascular aerobic exercise, pursued by millions and regularly included in New Year resolutions by many millions more, the *vicious circular* aspect of it is probably not so obvious and clear to us, but this is how Huxley represented it in the final chapter of his book.

This may be also viewed as one more prophetic clue by Huxley, alongside the genetic engineering or virtual reality entertainment of *Brave New World.* Shearwater's stationary bike ride with instruments to gauge different physiological indicators, presages our modern stationary bikes which calculate virtual speed, distance, calorie expenditure, wattage and other vital data:

> On the nightmare road he remained stationary. The pedals went round and round under his driving feet, the sweat ran off him. He was escaping and yet he was also drawing nearer.[6]

Forty years later, long after his rejection of scepticism and pyrronism and after becoming a devout pacifist and a "perennial philosopher," Huxley wrote his final novel *Island*, a "positive utopia," as it is usually known, to distinguish it from (his) "negative utopias" or "dystopias" – *Brave New World* and *Ape and Essence. Island* presents an "ideal" society living on the island of Pala in the Indian Ocean, a society based upon a happy blend of what Huxley considered to be the best in European empiricist philosophy, with various strands taken from Eastern religious and philosophical tradition.

As far as the Eastern tradition is concerned, most of the ideas in operation on Pala are taken from Buddhism, particularly Mahayana Buddhism. Buddhism was closest to Huxley's "Perennial Philosophy" because it is the least assertive in its attitude to a

4 Huxley, Antic Hay, p. 7.
5 Keith May, Aldous Huxley (London: Elek, 1972), p. 48.
6 Huxley, Antic Hay, p. 280-281.

personal God, Gods, and deities and closest to the ideas of an impersonal Godhead and/or the divine Ground, ideas which are pivotal in Huxley's "Perennial Philosophy." But in one of the crucial scenes in *Island*, the initiation into manhood (womanhood) ceremony at the very centre of it, we encounter a Hinduist God – Shiva.

Huxley's recipe for the initiation ceremonies for the young boys and girls is to first take them for some free-climbing and then give them four hundred milligrams of revelation, in other words "two firsthand experiences of reality, from which any reasonably intelligent boy or girl can derive a very good idea of what's what."[7] And while "the revelation's," the *moksha*-medicine's, the drug's effect lasts, the boys and girls look at the small representation, an image of Shiva-Nataraja, the Lord of the Dance: (Faulty syntax – rewrite)

> Look at his image ... look at it with these new eyes the *moksha*-medicine has given you. See how it breathes and pulses, how it grows out of brightness into brightness even more intense. Dancing through time and out of time, dancing everlastingly and in the eternal now. Dancing and dancing in all the worlds at once. ... In *all* the worlds. And first of all in the world of matter. Look at the great round halo, fringed with the symbols of fire, within which the god is dancing. It stands for Nature, for the world of mass and energy. Within it Shiva-Nataraja dances the dance of endless becoming and passing away. It's his *lila*, his cosmic play. Playing for the sake of playing, like a child. But the child is the Order of Things. His toys are galaxies, his playground is infinite space and between finger and finger every interval is a thousand million light-years. Look at him there on the altar. The image is man made, a little contraption of copper only four feet high. But Shiva-Nataraja fills the universe, is the universe...He dances because he dances, and the dancing is his *maha-sukha*, his infinite and eternal bliss ... .[8]

In *Island*, *moksha*-medicine, a hallucinogenic, mind-expanding drug, is to be used sparingly, once every year, to give people an hour or two of "enlightening and liberating grace"[9] so that they live full of attention, realizing all the time that they are part of that circular dance of Shiva-Nataraja. And Shiva becomes, in a nicely ironic way, a representation of that opening concept from *Antic Hay* of a "theotrope," "wheel of Gods," of monumental "theodrome." We have a stark contrast of two religious ceremonies or services for the young. In *Antic Hay*, Anglicanism, and more generally Christianity, are *passe*; a set of habits, rituals and wisdom that permits and helps construction of the playful concept of theodrome, a concept which is rejected as "nonsense" while the real remedy becomes pneumatic trousers. In the utopian *Island* Shiva-Nataraja as a theodrome, a circular, eternal

---

7 Aldous Huxley, Island (HarperCollins: New York, 2002), p. 189.
8 Huxley, Island, pp. 204-206
9 Huxley, Island, p. 208.

dance, remains a goal towards understanding which, towards the realization of which, all the Palean social arrangements, techniques and opportunities are directed.

To trace the utopian ideas of *Island* it is useful to return to the origin of some of the (strangely similar but) dystopian ideas of *Brave New World.* Throughout the 1920's and early 1930's, Huxley's sceptical and gloomy views on culture and civilization grew even more sceptical and gloomy. This process, discerned both in Huxley's fiction and in his non-fiction, is traced in detail in Robert S. Baker's *The Dark Historic Page: Social Satire and Historicism in the Novels of Aldous Huxley 1921-1939.* It is perhaps best summarized in the following:

> He came to regard the post-war society of the twenties and thirties as a fundamentally historicist culture, in which Hegelian notions of idealist history, Wellsian and Shelleyan ideas of historical progress, and what Huxley regarded as Marxist "modern" romanticism combined to create a bewildering ideological landscape, one that he traced back to the intellectual excesses of the romantic period. All of these thematic categories intersect in Huxley's exploration of ideological historicism, an analysis of a set of interrelated ideas such as "historical process," or more fundamentally, the prevailing tendency "to regard historicalness as a value."[10]

Although overtly debunking "the grand historical narratives," especially those with the ideas of historical, linear progress, Huxley occasionally returned to what he referred to as "historical undulations," wavelike pattern(s) of development, with "troughs" and "crests" or, as a character from *Point Counter Point*, Mark Rampion, saw it with "peaks" and "declines."[11] Most of Huxley's critics agree that *Brave New World* is a projection into the future of the trends and „undulations" that he saw in Europe and America at that time. The problem of history in *Brave New World* is solved in a simple way – Henry Ford's slogan "history is bunk" is adhered to, history is banished, and the historical process is halted. "Carefully controlled society involves an immersion in the present in which Pavlovian conditioning, Marxist collectivism, Fordean technology, and a calculated indulgence of Fruedian infantile appetitiveness combine to rigidly stabilize society and undermine the concept of linear progress."[12] The "here and now" of *Brave New World* is a sort of stupefying limbo, not the blissful eternity of the "mystic" *Island*, and the reality of it is to be escaped from, numbed by a daily dose of a drug called *soma*, which is, in turn, so different and yet also so ironically similar to the positive, mind expanding *moksha*-medicine of the utopian *Island.*

---

10 Robert S. Baker, The Dark Historic Page: Social Satire and Historicism in the Novels of Aldous Huxley 1921-1939 (Madison: University of Wisconsin Press, 1982), p. 5.
11 Baker, The Dark Historic Page, p. 137.
12 Baker, The Dark Historic Page, p. 139.

In order to discover some of the reasons for this radical shift in attitude to the problem of *here and now* and timelessness, let us now examine an essay on time that is part of Huxley's travel book *Beyond the Mexique Bay* (1933). The genre of the travel book, with its loose and episodic structure which encourages digressions, suited to Huxley's erudite essayistic style and the essay on time is just one of quite a few, but one which is particularly important from our time-oriented perspective. Visiting Copan – the ruins of ancient Maya culture – enables Huxley to discourse on time cycles Maya priests created and this fosters a series of more general remarks about the nature of time and human perception of it.

The temples, pyramids and stairways in Copan were erected and enlarged to celebrate the elapse of significant spans of time. The Mayas had a sacred "year" of 260 days, but they also used much longer cycles; Katun, a cycle of 7200 days and even 144000 days. It is also probable that they used larger units of 288000 days, and even of more than 1800 million days. Huxley believed that such a pre-occupation with time must have been the result of a series of personal accidents. A man is born to whom, for whatever reason, time is an obsession. He has the intellectual mastery to transform time into comprehensible quantitative terms. He is also in a position to influence others and make disciples. A tradition is formed, as was the case with Mayas; but in some other cultures, such as the ancient Greek, despite its plethora of accomplished mathematicians and philosophers, it was equally "natural" to ignore the subject.

But at the core of the problem is the psychological truth that time as such is unbearable.

> Any possible conception of time must be depressing. For any possible conception of time entails the recognition and intimate realization of the flux of perpetual perishing; and to be made aware of the flux – the flux in relation to one's own being; worse, as a treacherous and destructive element of that being – is intolerable. Regular, one, undifferentiated, time goes sliding beneath and through life, beneath and through its various pains and pleasures, its boredoms and enlightenments and seemingly timeless ecstasies – always the same mysterious lapse into nothing. The realization of it is, I repeat, intolerable. Not to be borne. ... Time is unbearable. To make it bearable, men transform it into something that is not time, something that has the qualities of space. [13]

Then Huxley discusses at some length the techniques we use to spatialize time; he lists nature with the marching of heavenly bodies, regular recurring of hunger, desire, sleep, days, nights, and seasons on to which men grafted all kinds of arbitrary systems of their own. Next he moves to artistic techniques "for parceling up the continuous flux" and "bending the irreversible flux into the semblance of a

13 Aldous Huxley, Complete Essays: Volume III, 1930-1935 (Chicago: Ivan R. Dee, 2001), pp. 556-557.

circle" and also "those biological and social devices for dulling men's awareness of the flux – habit and its social equivalent, routine."[14] He then tackles religion, which exploits the calendar with its feasts and ceremonies, as well as the time-transmuting arts of music, poetry and dance. And finally he moves to the philosophical and mystical concept that time is an illusion and eternity the only reality. In 1933, at the time of writing *Beyond Mexique Bay*, he does not accept this doctrine as true:

> But even if true – and personally I should like it to be true – the doctrine is not very effacacious against the obsessive consciousness of duration. For an illusion which is shared by all living beings, at any rate on our planet, is for all practical purposes indistinguishable from reality … . In the mind of the chronologist, the musician, the common creature of habit and routine, time has been transformed, by a variety of different processes into the likeness of a circle. The mystic goes one further and contracts the circle to a point. The whole of existence is reduced for him to *here, now.* Time has been spatialized to the extreme limit. But, alas, when he emerges from his ecstasy, he finds the current still flowing – realizes that it has been flowing even while he imagined that he had altogether abolished it. The flux may be an illusion, but it is an illusion always and unescapably there. [15]

As we have seen in the *Beyond the Mexique Bay*, Huxley's attitude to mysticism is sympathetic, but he clearly distances himself from it. In the quoted fragment, *he should like it to be true*, while in a different fragment when discussing the dangers inherent in the rise of nationalism in the early 1930's, he advises world educators to introduce „*mythology and a world view* which shall be as acceptable to the New Stupid as nationalism and as beneficial as the best of the transcendental religions."[16]

Huxley theoretized and psychologized human inability to bear the flux of time and the desire to convert it into space; this explanation also included the "mystic leap" he was to take later himself. Using more up-to-date terminology we could well explain Huxley's wrestlings with the notion of time in terms of *nostalgia.* As Svetlana Boym informs us:

> In a broader sense nostalgia is a rebellion against the modern idea of time, the time of history and progress. The nostalgic desires to obliterate history and turn it into private and collective mythology, to revisit time like space, refusing to surrender to the irreversability of time that plagues the human condition.[17]

That Huxley was, however, already undergoing a process of *mystical* conversion can be deduced implicitly from his later fiction. The conversion from Pyrronism to

14 Huxley, Complete Essays III, p. 558.
15 Huxley, Complete Essays III, p. 560.
16 Huxley, Complete Essays III, p. 502; my italics.
17 Svetlana Boym, The Future of Nostalgia (New York: Basic Books, 2001), p. XV.

mysticism is best described in fiction in the novel *Eyeless in Gaza* (1936). There another Huxley-like main character, Anthony Beavis, undergoes such a change as a result of a very long sequence of events, and one of the culminating stages of his conversion is meeting the first of Huxley's fictitious „perennial philosophers," Dr Miller, while trekking in Central America around 1933 (his origins could be traced to Dr Mac Phail from *Beyond the Mexique Bay*: "the head of the United Fruit Company's hospital ... (whose) professional reputation stands very high, but it is his kindness and his wisdom that have made of him the universal godfather of Guatemala"[18]).

The new Huxley was preaching pacifism at the social level, and became an active member of the Peace Pledge Union. He wrote, among other things, a collection of essays entitled *Ends and Means* (1937) warning in vain that it is impossible to achieve good ends on all levels, individual and social, while using bad means. On an individual level he became an advocate of the mystic eternal *here and now* against the temporal flux. This conviction is clearly depicted even in his titles, but also, to a much greater extent, in the themes of the books he was to write during the coming years.

In the novels *After Many a Summer Dies the Swan* (1939) and *Time Must Have a Stop* (1944) as well as in his non-fictional summa *The Perennial Philosophy* he began contrasting the pathetic attempts of unregenerate men and women to prolong their lives in fear-ridden temporal flux with the blissful contemplation of the eternal *here and now* experienced by his fictitious mystics John Propter and Bruno Rontini, as well as by the historical mystics and saints of all the religious traditions quoted and analysed in *The Perennial Philosophy*. The importance of the eternal, blissful *here and now* for Aldous Huxley after his „conversion" in the mid-1930's, was foregrounded much later by his second wife, Laura Archera, long after the death of Huxley. *This Timeless Moment* – that is how she entitled her book about their life together.

Summing up, it should be emphasized that in Aldous Huxley's writings, both in his fiction and non-fiction, we can detect a clear, 180- degree turn in his views and attitudes. Most central and most astonishing is the turnabout in his views on circularity in nature and culture. The young, flippant, witty, erudite Huxley toyed and juggled with many Western and Eastern philosophical ideas; among them was the mystical idea of the bliss of *here and now*, of the circularity of time, and of the convergence of this cycle to a single point. At that time it was just an idea like many others, to play with both philologically and ideologically, to have sympathy for but to reject it as a Pyrronist, an ultra sceptic, should reject all non-sceptical ideas. After his "conversion" in the mid-1930s it became, to use a circular metaphor, a hub in the wheel around which Huxley's "perennial philosophy" revolved.

18 Huxley, Complete Essays III, p. 471.

Sean Hartigan

Recycling (and Counter-Recycling) in William Blake's *The Marriage of Heaven and Hell*

William Blake's dislike of the spirit of mechanism is well-known; as he wrote in the prefatory poem to *Milton*, the factories of newly industrial England appeared to him as "dark Satanic mills," and a friend offering him a copy of the *Mechanics Magazine* in 1823 was told simply but with great energy, "These things we artists *hate*."[1] It seems probable, then, that Blake would have disliked having his artistic activities described as "recycling," a term borrowed from the language of industrial production. However, although Blake probably would not have used this word himself, the *idea* of recycling, of the reuse and transformation of cultural products, is a foundation of his work. Indeed, in the work of few other poets is recycling so naked and extensive. As has long been recognized, a main characteristic of Blake's work is its incorporation and use of an eclectic mix of religious, political, and literary ideas ranging from classical and Biblical tradition to the mysticism of Emmanuel Swedenborg and the theories of radical historians and antiquarians. "We have the same respect for Blake's philosophy," T. S. Eliot observed, "that we have for an ingenious piece of home-made furniture: we admire the man who has put it together out of the odds and ends about the house."[2] Eliot's remark should not blind us, however, to the fact that Blake was both serious and self-conscious in his practice of recycling, as he was in all matters of artistic theory and practice. At times Blake even places recycling at the heart of artistic production. In the first chapter of *Jerusalem*, for example, his mythic protagonist Los builds the City of Art, and in the sculptures that ornament the City are the raw materials for an eternity of recycling, since the themes and situations of human life are eternal and destined to be repeated forever. As Blake writes, "All things acted on Earth are seen in the bright Sculptures of / Los's Halls & every Age renews its powers from these Works / With every pathetic story possible to happen from Hate or / Wayward Love & every sorrow & distress is carved here / Every Affinity of Parents Marriages & Friendships are here / In all their various combinations wrought with wondrous Art / All that can happen to Man in his pilgrimage of seventy years."[3]

---

1 Quoted in Ruthven Todd, Tracks in the Snow: Studies in English Science and Art (London: The Grey Walls Press, 1946), p. 11.
2 Thomas Stearns Eliot, "Blake," in Thomas Stearns Eliot, The Sacred Wood: Essays on Poetry and Criticism (London: Methune, 1920), Bartleby.com, 1996, 21 Aug. 2006 <www.bartleby.com/200/>
3 William Blake, "Jerusalem," in The Complete Poetry and Prose of William Blake, Newly Revised Edition, ed. David V. Erdman (New York: Doubleday, 1988), p. 161.

This view of art as unending joyful repetition might seem optimistic if the idea of recycling were to imply entrapment and dreary repetition, as in a treadmill. However, Blake's work sometimes expresses similar concerns about recycling, similar fears of an intellectual world without freshness or originality. The early pamphlet "There Is No Natural Religion," for instance, openly acknowledges such fears of repetition and argues that they stem from profound human concerns, for "[t]he bounded is loathed by its possessor."[4] Echoing the treadmill metaphor, Blake argues that a mind or culture deprived of a continuous flow of fresh inspiration degenerates into fruitless mechanical complexity. "The same dull round even of a universe," he writes, "would soon become a mill with complicated wheels."[5]

For Blake, then, recycling has two possible meanings; on one hand, it can be an expression of creative renewal, on the other, a practice of mechanical sterility. Such a dual view, however, is quite characteristic of Blake. It suggests that the idea of recycling is implicated in the basic dualism of his philosophy, the struggle between opposing psychological or spiritual principles, one associated with passion, imagination, and vitality, and one with abstraction, rationality, and stasis. In his early work, these principles are present as the two "Contrary States of the Human Soul" in *Songs of Innocence and Experience*,[6] while in his later work, they are often expressed symbolically, most notably as characters in the mythological poems – in particular, Los, the Eternal Prophet, and Urizen, the Primeval Priest.

The relationship between these principles and the idea of recycling is perhaps most directly suggested, however, in *The Marriage of Heaven and Hell*, a book remarkable for being the only one of Blake's illuminated books to be written primarily in prose rather than poetry. While this is a significant fact in itself, as will be discussed later, it also makes this book particularly useful for the present study, as the prose passages in *The Marriage of Heaven and Hell* show more clearly and directly the influence of eighteenth-century cultural theory, opening a useful door for us into the complexities of Blake's system of thought. In this book, the opposed principles mentioned earlier are called Energy and Reason and are associated with two human types called the Prolific and the Devouring. These names suggest the importance of recycling to Blake, for the Prolific are producers, representatives of Energy and creativity, while the Devouring live by consumption, "receiv[ing]"[7] the creative work of the Prolific and, as Blake writes, transforming it into "chains"[8]

4 William Blake, "There Is No Natural Religion," in The Complete Poetry and Prose of William Blake, p. 2.
5 Blake, "There Is No Natural Religion," p. 2.
6 William Blake, "Songs of Innocence and Experience," in The Complete Poetry and Prose of William Blake, p. 7.
7 William Blake, "The Marriage of Heaven and Hell," in The Complete Poetry and Prose of William Blake, p. 40.
8 Blake, "Marriage," p. 40.

that bind sensual passions and creative energy. The spiritual conflict between these two types of people, who are and should be "enemies,"[9] as Blake declares, is thus carried out in part through a process of recycling in which the energetic work of the Prolific is appropriated and transformed into an instrument of psychological constriction and stasis.

Blake indicates the importance and profundity of this process in the historical narrative on Plate 11 of *The Marriage of Heaven and Hell,* a passage notable for its suggestion that Blake sees cultural recycling as the source of one of his great philosophical foes, moralistic religion. According to Blake, this type of religion originated in the earliest periods of human history, not in the sayings or writings of primitive shamans or priests, as we might expect, but rather with "the ancient Poets," who "animated all sensible objects with Gods or Geniuses."[10] In naming *poets* as the source of religious ideas, Blake invites us to take a historical approach to the interpretation of this narrative, for here he seems to invoke the commonly accepted eighteenth-century theory that all forms of human culture originated in ancient bardic culture, whose primary mode of expression was oral poetry. For instance, this is the view of the historian Adam Ferguson, author of the influential *Essay on the History of Civil Society* (1767). Surveying the accounts of classical writers like Herodotus and Strabo, Ferguson concludes that "[p]riests, statesmen, and philosophers, in the first ages of Greece, delivered their instructions in poetry, and mixed with the dealers in music and heroic fable."[11] Reasoning from an Enlightenment faith in a universal human nature, Ferguson extrapolates from the classical world to conclude that "every tale among rude nations is repeated in verse, and is made to take the form of a song. The early history of all nations is uniform in this particular."[12]

The orality of these ancient poets is important because of the eighteenth-century conviction that oral expression is innately more passionate than written expression. Indeed, this was regarded as the primary reason for the superiority of ancient poetry, an axiom of neoclassical literary theory. Thomas Blackwell, for instance, perhaps the most frequently cited of these theorists in his own time, saw his *Enquiry into the Life and Writings of Homer* (1736), not as a disinterested investigation, but rather as an effort to determine "By what Fate or Disposition of things it has happened, that None have equaled [Homer] in *Epic-Poetry* for two thousand seven hundred Years."[13] Although Blackwell and his contemporaries believed that there were many reasons why the ancient oral bards were superior to

---

9 Blake, "Marriage," p. 40.

10 Blake, "Marriage," p. 38.

11 Adam Ferguson, An Essay on the History of Civil Society, ed. Duncan Forbes (Edinburgh: Edinburgh UP, 1966), p. 173.

12 Ferguson, History of Civil Society, p. 173.

13 Thomas Blackwell, An Enquiry Into the Life and Writings of Homer (Hildesheim, NY: Georg Olms Verlag, 1976), p. 2.

modern poets, the most important was that the ancients were simple people who composed their songs spontaneously in the heat of passion, without the hindrance of intellectual rules and guidelines. This idea is memorably expressed by Blackwell, who explains that the nature of oral performance inspired the ancient poet with a near-divine madness, for "while his Fancy was warming, and his Words flowing; when he had fully entered into the *Measure*, was struck with the *Rhythmus*, and seized with the *Sound*; like a Torrent, he wou'd fill up the Hollows of the Work; the boldest Metaphors and glowing Figures wou'd come rushing upon him, and cast a *Fire* and *Grace* into the Composition, which no Criticism can ever supply."[14] Blake would have expected his eighteenth-century readers to be familiar with these ideas, and thus to understand that the "ancient Poets" of *The Marriage of Heaven and Hell* were figures of unsurpassed passion and imagination – in fact, clear embodiments of the energetic and creative type of human being he identifies as the Prolific. Blake's history of the ancient poets (and their downfall) should thus be read as an episode in the ageless conflict between the Prolific and Devouring classes of humanity.

The vitality of these Prolific poets can also be seen in the content of their poetry, which is equally important for our understanding of Blake's historical narrative. As he explains, the ancient poets "animated all sensible objects with Gods or Geniuses," creating in their poetry a cosmology in which all things are conceived of as living, speaking, singing human forms. The importance of this point is again best understood in historical context, for the opinion of eighteenth-century cultural theorists was that the animated cosmos of the ancient bards is the world-view that emerges spontaneously in a culture of unrestrained passion and imagination. The rhetorician Hugh Blair, for instance, argues in his *Lectures on Rhetoric and Belles Lettres* (1783) that "[i]n the infancy of all societies, men are much under the dominion of imagination and passion."[15] Under these conditions, "a wonderful proneness in human nature to animate all objects" leads to "the multiplication of divinities."[16] Blair notes the persistence of this habit in the rhetorical device of *prosopoeia*, or personification, but remarks that this device, while "the boldest of all rhetorical Figures," is not to be attempted except "when the mind is considerably heated and agitated … in a state of violent emotion."[17] This relation of personification to passion explains both its ubiquity in primitive cultures, and its importance in Blake's narrative of the origins of religion. This animated cosmos of pervasive human form is the habitat of the passionate and creative Prolific in their natural state, an Eden of imaginative freedom.

In *The Marriage of Heaven and Hell*, however, this Eden contains Devouring serpents who destroy bardic culture by recycling its key elements and

---

14 Blackwell, Life and Writings of Homer, p. 120.
15 Hugh Blair, Lectures on Rhetoric and Belles Lettres (New York: Garland, 1970), p. 141.
16 Blair, Lectures on Rhetoric, p. 409.
17 Blair, Lectures on Rhetoric, p. 409.

transforming them into instruments of mental constraint. As Blake writes, the creations of the ancient poets were appropriated by interlopers who "enslav'd the vulgar by attempting to realize or abstract the mental deities from their objects."[18] That is to say, the human forms used by the ancient poets to conceptualize sensible objects were separated from those objects and converted into abstract gods, distant and unreachable except through the medium of the interlopers themselves, who usurped the social position of the ancient poets and established themselves as a new class of priests. The passionate oral fables of the ancient poets became the raw material of a new, oppressive culture of Priesthood, and the dynamic figures that populated their poetic universe were recycled into powerful and vengeful gods ruling humanity through fear and moral law. The Priest is thus identified as a historic embodiment of the Devouring, transforming the creative work of the Prolific into the moral and legal "chains" that bind human Energy. "Thus began Priesthood," Blake concludes, "choosing forms of worship from poetic tales."[19]

This story illuminates one of the key principles of Blake's vision of recycling. The form of recycling practised by the Devouring is a process of *abstraction* and *reification*. This Blake sees as a powerful technique of mental war, a means of transforming humanity's experience of life by transforming the imaginative forms used to conceptualize that life. "The Eye altering alters all,"[20] as Blake writes in "The Mental Traveller," and to see the world through the Prolific eyes of the ancient poets is to experience "all alive the world, where every particle of dust breathes forth its joy,"[21] in the memorable phrase from *Europe: A Prophecy.* But when these myths are recycled and reified as a metaphysical system, humanity lives in fear of its own creations, believing them to have independent and absolute existence. "Thus men forgot," Blake writes, "that all Deities reside in the human breast."[22]

The purpose of the preceding discussion has been to show how Blake situates recycling at the heart of one of the great cultural transitions of human history, the decline of primitive bardic culture and its replacement by abstract, moralistic religion. But as an element of Blake's dualistic philosophy, recycling also has a positive dimension. *The Marriage of Heaven and Hell* suggests that in a post-bardic world dominated by the culture of the Devouring, the Prolific themselves may engage in a form of recycling. This Prolific recycling, practiced by Blake himself, is a process of *re-poeticizing* culture, a kind of *counter-recycling* in which the cosmos of the ancient poets is revived and the abstract products of

18 Blake, "Marriage," p. 38.
19 Blake, "Marriage," p. 38.
20 William Blake, "The Mental Traveller," in The Complete Poetry and Prose of William Blake, p. 485.
21 William Blake, "Europe: A Prophecy," in The Complete Poetry and Prose of William Blake, p. 60.
22 Blake, "Marriage," p. 38.

priestly culture are transformed into expressions of bardic energy. An example of this kind of transformation may be found in *The Marriage of Heaven and Hell*, which, as was noted earlier, follows a different format from those of Blake's other mature works, being composed primarily not of mythological poetry in the style of the ancient oral bards, but of philosophical prose, in the manner of the book-centered culture of Priesthood. This prose format is widely recognized as ironic and parodic, of the visionary Christian philosophy of Emanuel Swedenborg, particularly, but when the book is read in the terms of the historical narrative we have just examined, the nature of this parodic recycling is more clear: *The Marriage of Heaven and Hell* is the product of a bardic sensibility appropriating, recycling, and transforming the abstractions of priestly discourse. The very structure of the book suggests this, mirroring as it does the basic chronology of Blake's historical narrative. As we have seen, Blake believed that human history begins with the mythic culture of the ancient poets, only to fall into a period of mental slavery under the dominion of priestly culture and its moralistic theology. Similarly, *Marriage* opens with a mythological poem that helps to contextualize the parodic prose explorations of priestly Christian philosophy that follow. This poem expresses the wrath (given human form as Rintrah) of the usurped bardic culture, and tells a mythic version of its overthrow by the Devouring priests through the story of a "just man" who is driven into the wilderness by the "sneaking serpent"[23] who steals the beautiful garden he has planted. At the end of the book, following the prose sections, Blake seems to extend his historical narrative by including a second bardic poem, one that prophesies the end of priestly culture. This mythic "Song of Liberty" predicts the birth of Orc, Blake's poetic figure of revolutionary energy, and the collapse of the empires of the "sneaking serpents," the priests and their descendants. Their downfall is depicted metaphorically as a rejection and annihilation of the holy books that have replaced the songs of the ancient bards, as Orc "Spurning the clouds written with curses, stamps the stony law to dust."[24] As part of *Marriage*'s triumphant return to the poetic world-view of bardic culture, this passage manifests the very moment it symbolizes, even as *Marriage* itself manifests the psychological and cultural transformation it prophesies in the "Song of Liberty."

In the space between bardic culture's fall and rebirth, priestly culture thrives, and as noted above, the interior of *The Marriage of Heaven and Hell* is composed of mystical philosophy in the tradition of Swedenborg. But things are not exactly what they seem; the intellectual order of priestcraft shows signs of tampering. A foreign sensibility has infiltrated this order, stealing its outward forms and infusing them with the energetic spirit of bardic culture. Thus Angels (figures of abstract virtue in Swedenborg and in priestly Christian culture generally)

23 Blake, "Marriage," p. 33.
24 Blake, "Marriage," p. 45.

become stuffy prigs, while Devils (figures of the bodily energies that oppose this virtue) become swaggering heroes. An example of this adaptation can be found in the "Memorable Fancy" in which a rather irritable Angel and a Devil debate Jesus' relationship to the Ten Commandments. The Angel takes the orthodox position that Jesus "has ... given his sanction to the law of ten commandments," and that all other men are "fools, sinners, and nothings."[25] The Devil, on the other hand, interprets the Gospels according to his own conviction that "no virtue can exist without breaking these ten commandments,"[26] and contends that Jesus, when looked at rightly, in fact broke all of the commandments. Did he not mock the Sabbath, for instance, and so mock the Sabbath's God? "Jesus was all virtue," states this Devil, "and acted from impulse, not from rules."[27] On hearing this, the Angel abandons his moralistic perspective and becomes a Devil, and as Blake observes, "my particular friend."[28] The metaphysical universe of priestly theology is here recycled as comedy and its moral laws are replaced by incitements to disobedience. The figures of Angel and Devil, created by the ancient poets and reified by the culture of Priesthood, are recycled as poetic expressions of psychological perspectives, restored to their ancient status as portions of a mythic world-view.

The Marriage of Heaven and Hell thus presents us with a comprehensive vision of recycling as an agent of cultural and psychological transformation, a practice by which the imaginative Eye may be altered in the direction of either of the opposing principles of Blake's philosophy. More than this, the book itself is Blake's own act of counter-recycling, an attempt to reverse the ancient victory of the Devouring and inspire a renewed ascendancy of the Prolific spirit. This is the ultimate meaning of Blake's vision of counter-recycling. As he writes in "A Vision of the Last Judgment," his work is "an Endeavour to restore what the Ancients called the Golden Age,"[29] not universally and finally, for the Devouring, with their own tactics of recycling, are eternal – but in his own consciousness and in those of his readers.

---

25 Blake, "Marriage," p. 43.
26 Blake, "Marriage," p. 43.
27 Blake, "Marriage," p. 43.
28 Blake, "Marriage," p. 44.
29 William Blake, "A Vision of the Last Judgment," in The Complete Poetry and Prose of William Blake, p. 555.

PART II

BETWEEN HISTORY AND RELIGION

Aleksander Gomola

Does the Bible Say What It Says? "The Circular Dance" of Feminist Biblical Interpretation

Traditional and modern strategies of reading the Bible

The Bible, given its religious role and cultural status in the Western world, occupies an important place in hermeneutics. Read, interpreted, and translated even by non-believers,[1] the biblical text has been approached for thousands of years from practically every philosophical perspective within and without the community of Christians. Most interpretations have aimed at supporting and justifying the cultural, social or political *status quo* or a vision of the world; quite a few interpretations have used the Bible as a springboard to promote new ideas and new solutions to social or political problems. The former comprise biblical interpretations that helped to sustain slavery or a pre-Copernican vision of the universe in the past or creationist beliefs today; the latter comprise *sola scriptura* approach adopted by Luther or, to use a modern example, views proposed by modern feminist theology.

Biblical interpretation proposed by feminist theology may be an interesting object of hermeneutical study for a few reasons. Firstly, it may be seen as an attempt at adapting a seemingly totally inappropriate text to the specific needs and views of feminist theology; secondly it illustrates the interplay of ideas not necessarily in harmony in Christianity; thirdly, one may perceive it as a more or less successful solution to problems of the Christian community in the modern world.

There is no such a thing as one, homogenous feminist biblical interpretation. There are many schools within feminist theology and each of them approaches the Bible from its own, unique perspective. There is either a womanist interpretation of the Bible or a feminist one;[2] there are different hermeneutical schools such as the hermeneutics of recuperation, the hermeneutics of survival and the hermeneutics of suspicion.[3] There is also a postmodern feminist critique of the Biblical text represented by Mieke Bal and most of all by Luce Irigaray.[4] Among these various interpretations one of the most interesting seems to be the

1 An interesting example of a biblical translation by a non-believer into Polish is a translation of the Book of Genesis by Sandauer; see Artur Sandauer, Bóg, Szatan, Mesjasz i…? (Kraków, Wrocław: Wydawnictwo Literackie 1983).
2 George Aichele and Fred W. Burnett, The Postmodern Bible (New Haven and London: Yale University Press, 1995), pp. 225-244.
3 Aichele and Burnett, The Postmodern Bible, pp. 245-251.
4 Aichele and Burnett, The Postmodern Bible, pp. 254-258.

hermeneutics of suspicion adopted and promoted by a prominent feminist theologian, Elisabeth Schüssler Fiorenza. Elisabeth Schüssler Fiorenza, currently the Krister Stendahl Professor of Divinity at Harvard Divinity School, is one of the most important first generation feminist theologians and her works, especially concerning the Bible, exerted great influence on the development of this discipline in its beginnings. She has done pioneering work in feminist biblical interpretation presenting her ideas in many works on biblical hermeneutics. Her work has been recognized by some scholars[5] and criticized vehemently by others.[6]

Fiorenza's biblical interpretation goes beyond a few insights into this or that biblical passage, which is often the case with radical interpretations of the Bible promoted by various theological schools that do not belong to the mainstream of Christian thought. She has worked out a unique hermeneutic method of Biblical interpretation which aims at liberation of women from the enslavement that is the result of the patriarchal character of the Bible. Her "transformative dance of interpretation"[7] may have two variants. Both are discussed below followed by brief concluding remarks.

Feminist biblical hermeneutics as a "transformative dance of interpretation"

In one of her most seminal works the American feminist theologian proposes a unique strategy of reading the Bible, consisting of four movements: ideological suspicion, historical reconstruction, theoethical assessment, and creative imagination.[8] Fiorenza presents this strategy in a much more detailed way in another work, stressing that "these strategies of interpretation are not undertaken in a linear fashion. Rather, they must be understood as critical movements that are repeated again and again in the 'dance' of biblical interpretation" while "the reader is best imaged as a dancer, who engages in a circle-dance that spells not only movement but also embodiment."[9] What then are these movements and what is their role in feminist biblical interpretation as seen by Fiorenza?

The first movement of the "transformative dance of interpretation," namely ideological suspicion or "a hermeneutics of suspicion,"[10] prepares the ground for a proper, liberating interpretation of Bible. It rests on the presumption that "*all*

5 Richard Morgan and John Barton, Biblical Interpretation (Oxford: Oxford University Press, 1988), pp.157-160.
6 Garrett Green, Theology, Hermeneutics, and Imagination: The Crisis of Interpretation at the End of Modernity (Cambridge, England: Cambridge University Press, 2000), p. 203.
7 Elizabeth Schüssler Fiorenza, But She Said: Feminist Practices of Biblical Interpretation (Boston: Beacon Press, 1992), p. 53.
8 Elizabeth Schüssler Fiorenza, Bread not Stone: The Challenge of Feminist Biblical Interpretation (Boston: Beacon Press, 1983), p. 15.
9 Schüssler Fiorenza, But She Said, p. 52.
10 Schüssler Fiorenza, But She Said, p. 57.

biblical texts are articulated in grammatically masculine language – a language which is embedded in a patriarchal culture, religion, and society, and which is canonized, interpreted, and proclaimed by a long line of men."[11] Only after this is made clear is it possible to deconstruct the biblical text and to present its liberating character. The term "hermeneutics of suspicion," coined by Ricoeur to name the method of interpretation that goes back to the 19th century and whose forgotten father was Feuerbach,[12] is employed by Fiorenza to make the reader aware of the fact that a female reader and a male reader, if they read the same biblical text, are actually reading two different texts. A male reader reads the text that to a great extent conforms to his view of the world and social life (after all, it is a product of the patriarchal world), while a female reader must treat every biblical text with the greatest suspicion unless she wishes to accept the elimination of women from the world it presents or the reduction of their social status. A good example of this elimination is John 6:10, which reads: "Jesus said, 'Have the people sit down.' There was plenty of grass in that place, and the men sat down, about five thousand of them."[13] The only possible explanation for the discrepancy between the "people" who were ordered to "sit down" and the "men" who actually did so, is by presumption that the women and children who were also there are simply not mentioned by the biblical author. An example of reducing women's social status may be the constant addressing of Christian communities in Paul's letters by the word "brethren" although they consisted of both women and men. Fiorenza reminds the reader that we should not take such biblical texts at face value, especially those that deal directly with women.[14]

To avoid being trapped by the male centered text and its message, we should employ another strategy, i.e. "historical reconstruction" or a "hermeneutics of remembrance." Its aim is to "develop designs or models for historical reconstruction which can dislodge the eradicating frame of the androcentric biblical text."[15] The aim of this strategy is to restore women to their true place in biblical history, as agents that created and shaped it, only to be removed from it by the men who were writing biblical texts. As Fiorenza writes: "although women are neglected in the writing of history and theology, the effects of women's lives, thoughts, and struggles are a part of the historical reality and theological meaning that is concealed from us."[16] This "silent presence"[17] of women in Christianity throughout its history is clear evidence of the patriarchal character of the Christian

11 Schüssler Fiorenza, But She Said, p. 53.
12 Green, Theology, Hermeneutics, and Imagination, p. 83.
13 Holy Bible: New International Version (Zondervan, 1985), p. 431.
14 Schüssler Fiorenza, But She Said, p. 53.
15 Schüssler Fiorenza, But She Said, p. 54.
16 Schüssler Fiorenza, Discipleship of Equals: A Critical Feminist Ekklesia-logy of Liberation (New York: Crossroad, 1993), p. 266.
17 Elżbieta Adamiak, Milcząca obecność. O roli kobiety w Kościele (Warszawa: Więź, 1999), p. 12-15.

faith for centuries. A specific model of historical reconstruction of that presence proposed by Fiorenza is that of "discipleship of equals." Promoting this model, the American feminist theologian exposes the ignored role of women in creating and establishing the first communities of believers in the first decades of Christianity. Admitting that her theory is controversial to many theologians who regard it as "a feminist projection back into the first century, which has no support in our source texts," Fiorenza nevertheless insists that this support is possible to find if one uses a "hermeneutics of remembrance."[18] This strategy is similar to what archeologists do, trying to reconstruct the past from the broken pieces of a clay pot or, as Fiorenza writes, to the "activity of a quilt -maker who stitches all surviving historical patches together into a new overall design."[19] Because women were not allowed to voice their opinions directly, silenced by men, a modern reader should do it for them.

Expressing anew what was lost or removed from the message of early Christianity is possible thanks to the third movement of "dance of interpretation" which is the "theoethical assessment" or a "hermeneutics of proclamation." This movement seems to be the most important in the strategy of reading the biblical text proposed by Fiorenza, as it presupposes elimination of anything that does not serve liberation of women and other marginalized people (people of colour, slaves, servants, etc.). It is also highly controversial. "In theological terms, they [biblical texts] should not be proclaimed as the word of G-d but must be exposed as the words of men. Otherwise, Christian discourses about G-d continue ultimately to legitimize patriarchal oppression."[20] (Note that in order to eliminate the male connotations with the term "God," Fiorenza replaces it with "G-d"). In Fiorenza's own words "Like a health inspector, a hermeneutics of proclamation, for the sake of life and well-being, ethically evaluates and theologically assesses all canonical texts to determine how much they engender patriarchal oppression and/or empower us in the struggle for liberation."[21] These radical statements change altogether the role the biblical text has played in the communities of believers for hundreds of centuries. The Bible is no longer the ultimate source of reference or the source of moral obligation or social standards. What the Scripture says is accepted only as long as it corresponds with the principles adopted by the reader.

Finally the fourth, final movement of interpretive dance, "creative imagination" or "a hermeneutics of liberative vision" comes in and its aim is to "to actualize and dramatize biblical texts differently."[22] which in practice means a radical rewriting of the biblical text, adding, if necessary, something that should be present in the text or removing elements that are in contradiction with the values

---

18 Schüssler Fiorenza, Discipleship of Equals, p. 105.
19 Schüssler Fiorenza, But She Said, p. 54.
20 Schüssler Fiorenza, But She Said, p. 54.
21 Schüssler Fiorenza, But She Said, p. 54.
22 Schüssler Fiorenza, But She Said, p. 54.

and aims of feminist theology. This "creative re-imagination" may take not only literary form but may employ any other form of artistic creativity. It should also be an expression and manifestation of women's embodiment, their unique female experiences, sufferings, joys and insights. Embodiment as a specific female category is one of the distinguishing features of feminist theology. Along with the category of subjectivity it stands in contrast to spirituality and objectivity characteristic of traditional western philosophy and theology.[23] The "reconstructed" text may be used as an inspirational reading, read alongside the Bible by women, who find the biblical text oppressive. (For a sample of such a text, based on Luke 10: 38-42, see Appendix).

Feminist biblical hermeneutics in a global context

In one of her later works Fiorenza's presents an expanded version of her earlier hermeneutical strategy containing of the three more "steps": "a hermeneutics of experience that socially locates experience, a hermeneutics of domination, a hermeneutics of change and transformation." As in the case of the previous approach, "these strategies are not to be construed simply as successive and progressive but must be understood as corrective, repetitive and interactive."[24] That means that Fiorenza's process of biblical interpretation is still of a circular, not a linear, character, although the metaphor of a "dance" disappears from her language, replaced by totally different category. "If I were to identify a key interpretive metaphor for such a critical feminist approach (to the Bible – A.G), I would choose that of struggle."[25] Also her perspective of the existential situation of a female reader is much wider in her new book because the liberation of women and various forms of enslavement they have to struggle with must be seen in the global context.

Identifying globalization as one of the threats of traditional patterns of life and bearing in mind various conservative strategies of biblical interpretation adopted by fundamentalist groups within Christianity, Fiorenza calls on believers to rediscover the true meaning of one of the earliest Christian notions i.e. that of ekklēsia as the society based on the "discipleship of equals." This vision of the Christian community, along with principles of democracy, may help to create new and just global society. "Only the 'indigenization' of classical notions of democracy and biblical understandings of ekklēsia, ... with Western articulations of individual freedoms and equal rights, will result in a Christian vision and practice of radical egalitarianism that can fashion a global societal and religious

23 Susan Frank Parsons, ed., The Cambridge Companion to Feminist Theology (Cambridge, England: Cambridge University Press, 2002), pp. 43-87.
24 Elizabeth Schüssler Fiorenza, Sharing Her Word: Feminist Biblical Interpretation in Context (Boston: Beacon Press, 1998), pp. 76-77.
25 Schüssler Fiorenza, Sharing Her Word, p. 79.

vision of well-being for all without exception"[26]. Promoting this model of biblical interpretation, Fiorenza borrows two hermeneutical categories from another feminist theologian, Alicia Suskin Ostriker and speaks of a "hermeneutics of desire" and "a hermeneutics of indeterminacy." The latter strategy is the most important in a global context, according to Fiorenza, as it allows for its different interpretations of the biblical text and protects us from the rigid, fundamentalist and literal understanding of the Bible[27] that most often results in a "kyriarchal" pattern of social life and the enslavement of women.

Concluding remarks

The "transformative dance of biblical interpretation," the aim of which is to question the most popular interpretation of the biblical text and to call for its re-interpretation and re-construction, seems to be a very interesting, even if controversial to many, method of reading the Bible. Its controversy does not lie in the fact that the Bible is seen as a reflection of attitudes of a specific community of believers, as this approach goes back to the method of *Formcriticism* (*Formgeschichte*) initiated by Dibelius in early 20th century and is widely adopted in modern biblical studies today[28]. What is controversial to many is Fiorenza's insistence that the biblical text has no religious or moral value if it stands in contrast to the reader's vision of social and political relations in the world. In other words, we are not obliged to listen to the Bible if it collides with the preconceived ideas concerning what is just and right in a Christian community. This type of hermeneutics means shifting the centre of gravity within a religious community from doctrinal issues onto social and political ones.

Rewriting the biblical text is another controversial aspect of Fiorenza's hermeneutics. A tradition of relatively more or less innovative interpretation of the Bible is well known both in Judaism in the form of *midrash* and Christianity in the strategy of discovering of four senses of the biblical text defined by St. Thomas Aquinas,[29] although much older. Yet in the case of the hermeneutical strategy worked out by Fiorenza, the reader is encouraged to go beyond mere interpretation of the text and to change it to suit her/his goals and this is something most biblical scholars reject.

Fiorenza's hermeneutical approach to the Bible is part of a much wider debate concerning the place and role of the Scripture in modern Christian communities, both conservative and progressive ones. Whether it is the use of

26 Schüssler Fiorenza, Sharing Her Word, p. 133.
27 Schüssler Fiorenza, Sharing Her Word, p. 130.
28 Richard. N. Soulen, Handbook of Biblical Criticism (Atlanta: John Knox Press, 1981), p. 71-77.
29 St. Thomas Aquinas, Summa Theologiae, I, 1, 10 in Basic Writings of Saint Thomas Aquinas. Volume 1, ed. Anton C. Pegis (Indianapolis: Hackett Publishing, 1997), pp. 16-18.

inclusive language in biblical translation that caused controversies among American evangelicals,[30] the resurgence of creationism or a recent German feminist translation of the Bible where the biblical phrase "God the Father" is always rendered as "God the Father and Mother,"[31] the problem is the same: how to save the Bible and what to save from the Bible so that it may remain a defining element of a Christian community. Fiorenza's "dance of interpretation," although radical and controversial in many respects, is an interesting example of a hermeneutical strategy that aims at saving the Bible as the most important legacy of Christian communities, trying to proclaim at the same time its liberating force.

Appendix

Luke 10: 38-42 "re-created" according to the principles of Fiorenza's hermeneutical "dance of interpretation"

I am Martha the founder of the church in Bethany and the sister of Mary, the evangelist. All kinds of men are writing down the stories about Jesus but they don't get it right. Some use even our very own name to argue against women's leadership in the movement. Our great-great granddaughters need to know our true stories if the discipleship of equals is to continue. They had been travelling for a long time when they finally came to our village. I invited them to join my sister Mary and me. Jesus and the disciples with him sat down and began talking. Mary sat at the teacher's feet and I joined her in asking him about his latest journeys. He told us the story of the Syrophoenician who came to him asking that her daughter who was possessed be healed. Preoccupied with all the ministry to be done in Galilee Jesus refused: "I have come to serve only the lost and outcast of my own people." But to his great surprise -- Jesus continued -- the woman persisted and started to argue with him: God's gracious goodness is so abundant that the crumbs falling from Israel's table are sufficient for nourishing those who do not belong to God's special people. Her argument and faith was like a flash of revelation in which Jesus realized that the good news of liberation and God's power of wholeness was for all people, gentile and Jew, male and female, slave and free, poor and rich. And her daughter was healed. By the time the teacher finished this story, evening had approached and it was time for sharing the meal. I asked Jesus if he would stay to eat with us. He said yes, and added: "Martha don't go to a lot of trouble. Whatever you were going to have will be fine. Let me help you." We started toward the kitchen when one of the males hollered: "The women can go but you, Jesus, stay here. After all we have important things to talk about and they don't really

30 Donald A. Carson, The Inclusive Language Debate: A Plea for Realism (Michigan: Baker Books, 1998), pp. 22-40; Mark Strauss, Distorting Scripture: The Challenge of Bible Translation and Gender Accuracy (Downers Grove, Illinois: InterVarsity Press, 1998), pp. 13-17.

31 Bibel in gerechter sprache (Gütersloh: Gütersloher Verlagshaus 2006).

understand theology." But an Essene who had become one of the disciples travelling with Jesus said: "Isn't God's word for all people? Before I joined your movement I had always studied the Torah with other women. Are we women disciples to be excluded? After all, didn't your story about the woman from Syrophoenicia show that your message isn't just for some but for all, women and men, Gentile and Jew, slave and free, rich and poor?" And Jesus replied: "Susanna thank you for speaking out. You are much blessed by Holy Wisdom, for you are right." And he asked me to preside at the breaking of the bread and invited Susanna to say the blessing and to teach the Torah lesson for the day. There was grumbling among the men, but we women were excited by the new possibilities God had opened up to us.

My sister Mary helped me to write this down. May God Herself speak to us now and forever.[32]

32 Schüssler Fiorenza, But She Said, pp. 74-75.

Ewa Rychter

Precious Absence. Resurfacing Christianity in Gianni Vattimo and Slavoj Žižek

In 1951, Richard Niebuhr, Professor of Theology, wrote a book in which he focused on the "perennial" and "enduring" problem [1] of relations between Christianity and civilisation. The book bears the title *Christ and Culture*, and its subsequent chapters are discussions of five typical answers to the problem signalled in the title. The tentative types devised by Niebuhr, meant to systematise the multiplicity of answers supplied over time, include one type emphasising opposition between Christianity and civilisation ("Christ against culture"), one in which the two notions are in fundamental agreement ("Christ of Culture"), and three median types in various ways maintaining difference between the two principles ("Christ above culture," "Christ and culture in paradox," "Christ the Transformer of culture"). Niebuhr emphasises the inconclusive and open character of this typology – there may be other types; types can overlap as "strange family resemblances may be found along the scale"[2]; no type can be treated as "the Christian answer" to the Christianity-culture problem because the history of Christianity in culture is not concluded.[3] Niebuhr is careful not to usurp the privileged position of someone who recognises the objectively true type. Having denied himself the right to decide, however, he leaves no space for doubt that there *is* a transcendental answer to the recurrent problem of the relationship between Christianity and culture. "Christ's answer to the problem of human culture is one thing, Christian answers are another; yet his followers are assured that he uses their various works in accomplishing his own."[4] The imperfect, historically motivated, context-related answers[5] are orchestrated by Christ, who eludes all of Niebuhr's types and whose otherness in relation to those types allows him to employ all partial insights into his own absolute non-historical vision.[6] To use Niebuhr's

1 Richard H. Niebuhr, Christ and Culture (New York: Harper & Row, 1951), p. 2.
2 Niebuhr, Christ, p.40.
3 Niebuhr, Christ, p. 232.
4 Niebuhr, Christ, p.2.
5 Niebuhr does not deny the historicity of those answers, yet at a certain point he de-emphasises its significance, saying that "those partial answers [...] recur so often in different eras and societies that they seem to be less the product of historical conditioning than of the nature of the problem itself and the meanings of its terms." (Niebuhr, Christ, p. 40.)
6 For Niebuhr, Christ's essence cannot be described. Though you can only offer an interpretation of "phenomena in which his essence appears," such an interpretation should be "of the objective reality." (Niebuhr, Christ, p. 14.)

military metaphor, Christ is the captain who remains behind the battlefield and busily devises strategies.

In 2005, Graham Ward, Professor of Theology, wrote a book in which he focused on the enduring problem of Christianity and culture. The book also bears the title *Christ and Culture*, and its arguments rest on the premise that "Christ is already a cultural event," and that "we have no access to a Christ who has not already been encultured."[7] For Ward, Christ and culture are not sides of a binary, two disjunct phenomena relating to each other in various ways, but inseparable mutually dependent things. Just as there "can be no distillation of Christ from culture,"[8] there cannot be any statement about Christ that would not be "a statement about ourselves and the times and cultures we inhabit."[9] The Christ we enquire into is the Christ that reveals himself to us today – the contemporary Christ figured forth through contemporary culture, irreducible to the lost body of the past, placed in the line of two millenia of negotiations with his incarnation. Ward's metaphor for the Christ thus conceived is the Deleuzean schizo[10] – a mobile, liquid, deterritorialised subject passing through "relationships of intensities in a way that demands the surrender of the ego, of the subject-in-control."[11]

Far from being a simple polemic with Niebuhr's *Christ and Culture*, Ward's *Christ and Culture* returns to and revisions the problem discussed by the classic of 1951. Significantly, in his return Ward reinscribes the fundamentals of the focal problem, collapsing the two terms Niebuhr kept separate, weakening the figure of Christ (compare the captain with the schizo), making fluidity and flow rather than typology the key to the problem of Christ and his presence in culture. Ward's schizoid Christ, however, neither replaces Christ the captain from Niebuhr's book nor constitutes the long-awaited, right answer to the titular problem. Rather, the schizo is what remains of the captain whose battlefield mastery and power are abandoned but whose memories of mobility and dissipation are left. In a revealing metaphorical description, Ward makes his Christ "surrender" his ego, leaving us to speculate about the schizo as an abased captain emptied of his war expertise and wandering about unconcerned with strategies or ahistorical perfection. Admittedly, Ward neither corrects not neglects the imperfections of Niebuhr's argument. Even though he avers that Christ and culture are more

7 Graham Ward, Christ and Culture (Oxford: Blackwell, 2005), p. 21.
8 Ward, Christ, p. 21.
9 Ward, Christ, p. 2.
10 "As for the schizo, continually wandering about, migrating here, there, and everywhere as best he can, he plunges further and further into the realm of deterritorialisation, reaching the furthest limits of the decomposition of the socius on the surface of his own body without organs. ... The schizophrenic deliberately seeks out the very limits of capitalism: he is its inherent tendency brought to fulfilment, its surplus product, its proletariat, its exterminating angel. He scrambles the codes and is the transmitter of decoded flows of desire. ... Schizophrenia is desiring-production as the limit of social production." (Ward, Christ, p. 35)
11 Ward, Christ, p. 61.

symbiotic than Niebuhr could ever acknowledge, he retains the title of Niebuhr's book, whose and-structure is the visible indicator of the separability of Christ and culture. Curiously, Ward's publication – which does not mention Niebuhr, except for one meagre, half-page appearance – keeps the not symbiotic enough and-structure of the title. Thus, "Christ *and* Culture" maintains a peculiar relationship with the other work from which it differs in nearly all respects. Emptied of the meanings it soaked up in Niebuhr, but never refashioned to erase its past, the title "Christ and Culture" encodes both impotence and persistence. It designates a problem or a challenge that seems to depend on and return into the precarious space of titles – the confusing, vaguely demarcated *front*-line of a text.

My point in this rather longish presentation of the Christ-and-culture theme twice-born is that the way Ward returns to the problem title/title problem is symptomatic of the way Gianni Vattimo and Slavoj Žižek turn to Christianity and culture in their recent writings. What those two philosophers seem to share with Ward – the differences between them notwithstanding – is on the one hand, the "enculturing" and weakening of Christ, and on the other, moving their impotent and imperfect Christianity to the surface of contemporary culture. In the hands of philosophers unconstrained by the demands of theology, such an "encultured" Christ and a paratextualised Christianity offer a perplexing insight into the problems of change, return and fidelity to an inaugural event of the Western culture. In what follows, I will sketch the most important points at which Žižek's and Vattimo's voices – far from engaged in an *unisono* performance – speak with the same timbre.

It is quite striking that over recent years both Vattimo and Žižek have relocated the problem of Christianity from the level of important but submerged issue to the level of titles and main arguments. In Vattimo's *The End of Modernity* (1988) and *The Adventure of Difference* (1993), Christianity appears as a modest companion of Nietzsche and his death-of-God statement. Some changes can be observed in *Beyond Interpretation* of 1996: although Vattimo places the Christian religious tradition under the general heading "Religion" and thinks it as important as the issues of science, ethics and art (considered in other chapters of the book), he makes Christianity an eye-catching rather than merely implicated part of the broader hermeneutical problematic. Finally, in *Belief* (1999) and *After Christianity* (2002), Vattimo's whole argument starts to revolve around Christianity. Similarly, in many of Žižek's writings published before 2000, Christianity functiones as a source of memorable analogies for central problems. Thus, in *The Sublime Object of Ideology* (1989), *che vuoi?* is discussed in the context of Jewish and Christian faith; in *The Plague of Fantasies* (1998), "subject supposed to believe" and "subject supposed to know" are illustrated by means of the Turin Shroud controversy; in *The Ticklish Subject* (1999), grace remains related to the problem of free assumption of one's imposed destiny; in *For They Know Not What They Do* (1991), Saint Paul's legacy is presented as crucial for materialist concerns. In

Žižek's more recent work, however, Christianity has become an un-a-*void*-able, out-standing point on his agenda, splashed across title pages of his books (*The Fragile Absolute, Or Why is the Christian Legacy Worth Fighting For?* and *The Puppet and the Dwarf. The Perverse Core of Christianity*).

Apart from this shared relocation of Christianity, Vattimo and Žižek – each in his own way – define Christ(ianity) through weakness or imperfection, through absence of secrecy, through revelation as the absence of mystery. In his *Belief* and *After Christianity*, Vattimo – an Italian postmodern philosopher – sees the incarnation (synonymous with kenosis) as the constitutive event not only for the Christian religion but also for Western culture. Theologically, kenosis means the emptying out by means of which St Paul described Jesus as incarnate God. Vattimo's interpretation of St Paul's *kenoō* – an interpretation showing some affinity with contemporary ethical readings of the fragment from the Letter to Philippians 2.5-11[12] – assumes that the "emptying out" referred to by Paul stands for the abasement, weakening and diminishing of the divinity who comes to inhabit a human body. Kenotic Jesus is a weakened, non-violent divinity, different from the omnipotent, impenetrable, threatening deity of natural religion. Incarnation, in other words, brings an end to the sacred as violence and to the scapegoat mechanism on which sacred violence relies.[13] For Vattimo, kenosis amounts to "the

---

12 In modern New Testament studies, there are two main strains of interpreting Paul's kenosis. The first one is the so called the "gnostic redeemer theory," according to which Paul's fragment uses some pre-Christian material featuring an archetypal original Man descending to the earth to impart secret knowledge. Borrowing from the pre-Christian hymn, Paul writes of Christ as the redeemer who existed before his descent to the earth, and whose appearance, i.e., kenosis, is part of "the mythological rhythm of salvific intervention and release." (Sarah Coakley, Powers and Submissions: Spirituality, Gender and Philosophy (Oxford: Blackwell, 2002), p. 7). The second interpretation is the ethical reading within which kenosis connotes Christ's earthly servant-like attitude and has nothing to do with the problem of pre-existence. It is maintained that in the Philippians fragment Paul develops his Adam typology, in which kenosis stands for Christ's undoing of Adam disobedience (Adam – like Christ – was in the form of God, but he – unlike Christ – decided to be even more like God and ate from the Tree of Knowledge). Though in the ethical reading (contrary to Vattimo) kenosis does not denote incarnation, Vattimo's kenosis seems to be indebted to the ethical reading insofar as he withdraws from the strictly metaphysical considerations characteristic for the dogmatic reading (explained below). Moreover, Vattimo's temporality substantially weakens the problem of pre-existence (the Beyond), which to some extent dovetails with the contemporary anti-metaphysical attitude to Paul's kenosis. I think it is quite revealing (and supportive of Vattimo's weak thought) that today's New Testament discussions unanimously rule out the so-called dogmatic interpretation of Paul's kenosis. The dogmatic reading emerged as an effect of christological discussions in the fourth and fifth centuries and presupposed Christ's substantial pre-existence and essential divinity. For example, for Cyril of Alexandria, a theologian of the fifth century, kenosis did not entail any loss of divine features but was associated with an abasement stemming from the taking of flesh. Cyril's incarnated Christ was not stripped of any aspects of divinity he possessed as the Logos before the event of incarnation: his divine powers were not restricted but appended with the addition of human flesh.

13 Gianni Vattimo, Belief, trans. Luca D'Isanto (Cambridge: Polity Press, 1999), p. 37.

removal of all the transcendent, incomprehensible, mysterious and even bizarre features"[14] that for so long have kept reason and faith separate. Through kenosis, argues Vattimo, the idea of God as radically other and beyond the powers of thought gives way to the idea of God as a friend,[15] an unmasker[16] of myths, an interpreter who reads the Scriptures in a dialogue with the specific context and remains aware of the inconclusive character of his unveiling of prophesies.[17] Jesus the Supreme Interpreter, withdrawing to make room for the Holy Spirit and for more and more spiritual (non-literal) interpretation, shows that "salvation takes place through interpretation"[18] rather than through the acceptance of the given, unchangeable truth. The interpretative path of salvation is coexistent with secularisation which Vattimo understands as "desacralising the violent,"[19] as "the progressive dissolution of the natural sacred," and as "the very essence of Christianity."[20] If incarnation is "an archetypical occurrence of secularisation,"[21] salvation can enfold only within the process of secularisation, which for Vattimo – unlike for the majority of theorists of secularisation[22] – preserves and reproduces the continuity of Christian message.

What does the kenotic "dissolution of divine transcendence"[23] mean for Vattimo's philosophical reflection? Translated into philosophical discourse, incarnation/kenosis emerges as weak thought. As Vattimo divulges, weak thought relies on "the idea that history of Being has as a guiding thread the weakening of

---

14 Vattimo, Belief, p. 55,
15 Vattimo, Belief, pp. 49, 53.
16 Vattimo, Belief, p. 66.
17 Vattimo, Belief, p. 79.
18 Vattimo, Belief, p. 60.
19 Vattimo, Belief, p. 42.
20 Vattimo, Belief, p. 50.
21 Gianni Vattimo, After Christianity, trans. Luca D'Isanto (New York: Columbia University Press, 2002), p. 67.
22 It is quite difficult to identify Vattimo with any of the sides of the secularisation debate. His understanding of secularisation to some extent resonates with the so called "Durkheimian thought," (Judith Fox, "Secularisation," in: John R. Hinnells, ed., The Routledge Companion to the Study of Religion (London: Tylor and Francis, 2005), p. 294.) within which the decline of the social significance of Christianity is not equated with the fact that Christianity itself has become irrelevant (e.g., Grace Davie's "believing without belonging" thesis [Grace Davie, Religion in Britain Since 1945: Believing Without Belonging (Oxford: Blackwell Publishing, 1994.]). In Vattimo, there are also "echoes" of the so called "Weberian" heritage (Fox, "Secularisation," p. 294) visible in such claims as Steve Bruce's idea that religion's return today is not the reason for but the effect of important political and social changes (Steve Bruce, God Is Dead: Secularization in the West (Oxford: Blackwell Publishers, 2002), p. 31). Moreover, Vattimo's non-metaphysical approach to secularisation dovetails with claims which describe secularisation as a narrative rather than a scientific theory (Jeffrey Cox, "Master Narratives of Long-Term Change," in: The Decline of Christendom in Western Europe, 1750-2000, Hugh Mcleod and Werner Ustrof, eds. (Cambridge: Cambridge University Press, 2003), p. 206).
23 Vattimo, After Christianity, p. 27.

strong structures."[24] Weakening is "the constitutive character of being in the epoch of the end of metaphysics,"[25] i.e., at the time of the exhaustion of the belief in an objective world order which thinking should contemplate and accept. Weak thought is a reflection on the "long farewell to the strong structures of Being," a farewell "conceived as an indefinite process of consummation and dissolution of the structures themselves."[26] The process is "indefinite" since the violence of metaphysics and the logic of sacrifice cannot be effectively overcome but must be continuously reduced anew.[27] Not to perpetrate the violence of strong structures, the reflection carried out by weak thought should display a hermeneutical character in that it should not attempt to grasp the "original" meaning or the "objective" presence of the clearly defined things, but see itself as functioning *inside* "a diminished reality,"[28] *within* the play of interpretations. Weak thought *is an interpretation*, just as "Jesus is an interpretation, an event of interpretation."[29] Weak thought grows out of and motivates the hermeneutic mission of Jesus, which means that on the one hand, weak thought derives from the event of incarnation insofar as the incarnation inaugurates the slackening of metaphysical thinking visible in Jesus' exegesis of Scriptures; on the other hand, weak thought as a postmetaphysical phenomenon allows us to see the incarnation as the inaugural and today still meaningful "hermeneutical" event.[30]

Admittedly, there is an interesting "effect" of making postmetaphysical thought the closest ally of the incarnation (by "effect" I mean here part of the hermeneutic circle in which weak thought and kenosis condition each other, and not simple linear consequence). The "effect" is expressed in Vattimo's belief that "faithfulness to the gospel should be sought in Nietzsche, especially in his announcement of the death of God."[31] Vattimo – neither the first nor the only one who reads Nietzsche as a Christian apostle – explains the God-is-dead aphorism as a transformative announcement of the possibility of responding to Being in a different, non-metaphysical way. The death Nietzsche announces means that "there is no 'objective,' ontological truth that might be upheld as anything other than friendship, will to power or subjective bond."[32] But beyond that, Vattimo interprets (weakens?) Nietzsche's thought as partaking of kenotic history within which it repeats the crucifixion and compehends the death of Jesus narrated in the Gospels. Christ's dissolution of metaphysics into unapodicting friendship and charity as the

24 Vattimo, Belief, p. 35.
25 Vattimo, Belief, p. 35.
26 Vattimo, Belief, p. 63.
27 Vattimo, After Christianity, p. 119.
28 Vattimo, After Christianity, p. 50.
29 Vattimo, After Christianity, p. 59.
30 Vattimo, After Christianity, p. 67.
31 Vattimo, After Christianity, p. 104.
32 Vattimo, After Christianity, p. 105.

new condition of truth not only resonates with the famous aphorism but also comes to be acknowledged and re-articulated in Nietzsche.

Let us leave Vattimo's Jesus the unmasker and take a close look at Slavoj Žižek and his *Christ-avec-Lacan* approach to Christianity as discussed in his *On Belief*, his *The Fragile Absolute, Or Why is the Christian Legacy Worth Fighting For?* and *The Puppet and the Dwarf. The Perverse Core of Christianity*. Reflecting on the exceptional status of the Christian religion, Žižek – in his usual bold manner – claims that "what is revealed in Christianity is not just the entire content, but, more specifically, that *there is nothing – no secret – behind it to be revealed.*"[33] Unlike in Judaism or in the Gnostic reinscription of Christianity (which rely on the ineffable resisting revelation or on a mystery accessible to the select few), in the Christian religion "there is no place for any God of Beyond."[34] Through Christ's incarnation and death, Christianity reveals the secret to which the Jews remained faithful – the secret of God's impotence. How is this disclosure accomplished? First of all, the incarnated Christ renders visible the horrifying secret that motivates the Jewish iconoclastic prohibition. According to Žižek, Judaism prohibits depicting God "not because an image would 'humanise' the purely spiritual Entity, but because it would render it all too faithfully" as "another PERSON in the fullest sense of that term."[35] While the Jewish prohibition is supposed to keep secret the antropomorphisation or humanity of God, Christ's incarnation brings back to the surface that which God has always secretly been: a person experiencing "full wrath, revengefulness, jealousy, etc., as every human being."[36] Christ as "JUST ANOTHER HUMAN BEING, a miserable man indiscernible from other humans with regard to his intrinsic properties"[37] is a proper revelation of the divine secret.

Significantly, Žižek does not make Christ *simply* human but, acknowledging his exceptional status, asserts that Christ is both "fully human" (remember Pilates's "*ecce homo*")[38] and "not fully human" (he is God, after all).[39] He is not fully human because he is marked with divinity conceived as a fleeting appearance, as a brittle, precious, "easily corroded"[40] dimension shining through the ordinary. Žižek calls that divinity a "fragile Absolute" which must be handled "as carefully as a butterfly,"[41] "a grimace,"[42] "a kind of obstacle, ... a 'bone in the

33 Slavoj Žižek, The Puppet and the Dwarf. The Perverse Core of Christianity (Cambridge, Massachusetts: The MIT Press, 2003), p. 127.
34 Slavoj Žižek, On Belief (London: Routledge, 2001), p. 91.
35 Žižek, On Belief, p. 129.
36 Žižek, On Belief, p. 129.
37 Žižek, On Belief, p. 131.
38 Žižek, On Belief, p. 90.
39 Žižek, The Puppet, p. 80.
40 Slavoj Žižek. The Fragile Absolute: Or, Why is the Christian Legacy Worth Fighting For? (London: Verso, 2001), p.128.
41 Žižek. The Fragile Absolute, p. 128.
42 Žižek. The Fragile Absolute, p. 104.

throat' – ... something, that unfathomable X ...,"[43] "the rupture/gap."[44] The Absolute is the barely nothing on the surface of Christ's humanity, the excess disturbing his human shape. Such weak divinity, suffering as it were from its own incomplete, gapped status, characterises Christ and reveals the character of divinity as such. Christ's anguished cry from the cross " My God, why have you forsaken me?" reveals the gap separating God (Father) from God (Christ), i.e., it reveals the gap *in* God. In Christ, the enigma *of* God is revealed as enigma *in* God and *for* God himself as God stumbles across the limit of his own omnipotence. As Žižek argues, God's impotence and his helplessness uncovered at the crucifixion bring death to God-the-Father.[45] Yet this self-abandonment opens the only available way for humans to identify with God. In Žižek's words,

> When I, a human being, experience myself as cut off from God, at that very moment of the utmost abjection, I am absolutely close to God, since I find myself in the position of the abandoned Christ. There is no "direct" identification with (or approach to) the divine majesty: I identify myself with God only through identifying myself with the unique figure of God-the-Son abandoned by God. ... Man's identity with God is asserted only in/through God's radical self-abandonment, when his distance towards God overlaps with the inner distance of God towards himself.[46]

The wound opening for humans in Christ/God figures not only as the revealed secret of divinity but also as the badge of Christ's full humanity. If it is true – and Žižek asserts it is – that to be human means to suffer from self-abandonment, to experience both a distance within oneself and an excess that cannot be fully integrated into one's life, it follows that Christ is fully human. Žižek comments on this paradox of Christ's (in)humanity in the following way,

> Christ is fully a man only insofar as he takes upon himself the excess/ remainder, the "too much" on account of which a man, precisely, is never fully a man: his formula is not Man=God, but man=man, where the divine dimension intervenes only as that "something" which prevents man from attaining his full identity.[47]

What is ultimately most important about Christianity is the fact that it never comes to terms with the confessed secret and the revealed excess. Christianity does not reintegrate the divine/human rupture into anything resembling the scheme of cosmic reconciliation known from the pagan cosmology of balance. Christ's death

43 Žižek, On Belief, p. 89.
44 Žižek, The Puppet, p. 80.
45 Žižek, The Puppet, p. 126.
46 Žižek, On Belief, p. 145.
47 Žižek, On Belief, pp. 130-131.

is neither fitted into a cycle of disruption and rebirth of divine order, nor treated as a catastrophic derailment whose violence can be attenuated through symbolisation. As Žižek argues in *The Fragile Absolute*, "Christianity asserts as the highest act precisely what pagan wisdom condemns as the source of Evil: the gesture of *separation*, of drawing the line, of clinging to an element that disturbs the balance. ... Christianity ... *is* the violent intrusion of Difference that precisely throws the balanced circuit of the universe off the rails."[48] As an example of such violent intrusion into anaemic reality, Žižek evokes Christ's words about turning the other cheek – words which interrupt the circular mechanism of restoring balance. To turn the other cheek rather than return the blow is an act that cancels the logic of re-establishing equilibrium. In fact, the "cheek" example brings together several threads of Žižek's argument. First of all, it condenses the nuances of Žižkean reflection on Christian fragility, violence and love: Christ the *fragile* God disposing of the logic of retribution to replace it with love-your-enemies injunction, does not reintegrate the trauma into the Law but *cuts into* "the Gordian knot of the vicious cycle of Law and its founding transgression."[49] But beyond this, Christ turning the other cheek makes "an unheard-of gesture of leaving behind the domain of the Law itself, of 'dying to the Law,' as Saint Paul put it ... ."[50] The theme of dying to the Law is developed by Žižek into the claim that Christianity is able to eliminate what he calls "the spectral obscene supplement of the Law" – the perverse core – and to "unplug" or "uncouple" true love.

The perverse core is an obscene secret Christianity is haunted with, the spectre Christianity struggles to reveal. The obscene supplement is the hidden companion of the prohibitive Law, a companion which creates the transgression against which the prohibition must be formulated. The Law produces transgression and makes it all the more desirable, which – as Žižek reminds us – was what Paul explored in the passage about Law engendering sin, i.e., Law producing the desire to transgress the Law.[51] Though the Law and its obscene supplement remain tightly related, their relationship can be brought to light when we examine the words in the "Do not kill!" or "Do not sin!" commandment. Taken as a unified sentence, the commandment conceals its spectral double; separated into the pure prohibition "Do not!" and the obscene order "Kill!" or "Sin!", the suspension of moral obligation accompanying the obligation itself is revealed. In the Christian context, the obscene supplement functions as "the secret initiatic knowledge"[52] and motivates perverse

48 Žižek. The Fragile Absolute, p. 121.
49 Žižek. The Fragile Absolute, p. 100.
50 Žižek. The Fragile Absolute, p. 100.
51 The biblical passage alluded to here is the fragment from Matthew 5:39 ("But I say unto you, That ye resist not evil: but whosoever shall smite thee on thy right cheek, turn to him the other also"), in which Jesus offers an alternative to the Hebrew Law (the one recorded in Exodus 21:24 and Leviticus 24:20) based on the retributive rule "an eye for an eye, and a tooth for a tooth."
52 Žižek, On Belief, p. 141.

readings of Christ's sacrifice. Thus, because Christ paid the price for all our transgressions, we are free to indulge in our desires without paying the price ourselves. As a result, we respond to the obscene call "Enjoy!" and make our life a version of a perverse prayer to the Virgin Mary, which Žižek playfully refers to: "O thou who conceived without having sinned, let me sin without having to conceive."[53] In another perverse reading of Christ's death, "God first threw humanity in order to create the opportunity for saving it through Christ's sacrifice."[54] Applied to the Judas story, the perverse reading will see Christ as covertly encouraging Judas to betray him and help him in that way to carry on the work of salvation.[55]

For all the seductiveness and contemporary appeal of such perverse Christianity, the really interesting and subversive potential of Christianity lies in its ability to leave the perverse behind. This can occur through "dying to the Law," i.e., through "unplugging" (separation) from the established symbolic rules, through "uncoupling" from the symbolic Law and its spectral supplement,[56] all of which is possible in Christian love. Christian love, whose structure differs from that of desire or of idolising, does not depend on secrets but revolves around appearances. Love trusts appearances which it never opposes to reality. Unplugged from the logic of "this is not that" or of "A is in fact B," love "FULLY ACCEPTS that 'this IS that,' ... that Christ, this wretched man, IS the living God."[57] Alluding to the famous love hymn from 1 Corinthians, Žižek claims that love believes everything[58] – believes without succumbing to the cynicism of the "suspended belief" (also called "disowned/impersonal belief"[59]), which makes us practice things we do not really believe in, and which "can thrive only as not fully (publicly) admitted, as a private obscene secret."[60] Love loves, says Žižek, because of the loved person's imperfections, not despite them. Love is that which continues even though everything else ends – it is the remainder, something that sticks out, tearing the cloak of totality. The work of love – with its repeated effort of

53 Žižek, The Puppet, p. 49.
54 Žižek, The Puppet, p. 53.
55 Žižek, The Puppet, p. 16.
56 Žižek. The Fragile Absolute, p. 130.
57 Žižek, On Belief, p. 90.
58 "Charity suffereth long, and is kind; charity envieth not; charity vaunteth not itself, is not puffed up, Doth not behave itself unseemly, seeketh not her own, is not easily provoked, thinketh no evil; Rejoiceth not in iniquity, but rejoiceth in the truth; Beareth all things, believeth all things, hopeth all things, endureth all things. Charity never faileth: but whether there be prophecies, they shall fail; whether there be tongues, they shall cease; whether there be knowledge, it shall vanish away. For we know in part, and we prophesy in part. But when that which is perfect is come, then that which is in part shall be done away." (1 Corinthians 13:4-10) The fragment in question seems to reverberate not only in Žižek's writing on love but also through some of his reflections concerning Redemption.
59 Žižek, The Puppet, p. 8.
60 Žižek, The Puppet, p. 6.

unplugging, with its abandoned quest for perfection and its Christ-like fragility – is the good news brought by Christianity.

What does the good news mean for Žižek's philosophical reflection? Christian love as the renunciation of correcting imperfections and as resistance to the deceptive totality (did Paul not say that love never ends?) is the excess in the Christian confession/revelation logic. The ruptured character of Christian confession/revelation, which transpires in love, sheds new light on the relationship between psychoanalysis and Christianity. Žižek's formula of Christian untotalisable revelation locates psychoanalysis in the nearest vicinity of thus-conceived Christianity and challenges the thesis about the non-Christian character of psychoanalysis. In other words, psychoanalysis defined through its acceptance of something beyond redemption/deliverance proves akin to Christianity defined through the unintegrable remainder. The feature of psychoanalysis which Geoffrey Hartman dubbed some time ago as "kakangelic" (meaning bad news[61]) and which refers to the presence of the incurable, unredeemable kernel, for Žižek dovetails with the most precious Christian legacy, with its evangelical good news of untotalisable work of love.

There is one more reason for which Christian love looms large in Žižek's recent writing. Love uncoupling from the Law is read by Žižek as the model of revolutionary change badly needed, he asserts, in times of complaisance, political correctness and cynicism. The revolutionary change – in Žižkean, the act – brings about the new beginning and enables us to reinvent ourselves. Žižek calls the act the impossible that happens, a "miracle" [62] that can be reflected on only retroactively, "a passive decision" which can "change Eternity itself."[63] The act as the "founding gesture of the primordial choice"[64] is "the Original Sin itself, the abyssal disturbance of primeval Peace ... ."[65] Žižek emphasises that even though the act connotes choice as sin or even as betrayal, it nevertheless has salutary

61 In "The Interpreter's Freud," Hartman argues that Freud's psychoanalysis offers a hermeneutics whose power can rival that of the biblical religions. Freud focuses on the usually repressed and sacred phenomena and offers a new textuality into which he introduces the so far excluded. Moreover, Freud re-opens the extreme indeterminacy already submerged in the Bible under the weight of authoritative commentaries, and exercises the cross-fertilisation between the text and its commentary, no longer available to us in the case of the Scripture. Another "feature that distinguishes psychoanalytic interpretation is its kakangelic rather than evangelic nature. I admit to coining this discordant word. The New Testament claims to bring good news, and reinterprets the Old Testament - that is, the Hebrew Bible - in the light of its faith. If the Gospels emphasize mankind's guilt, they also counter it by the possibility of salvation. But Freud brings bad (kaka) news about the psyche, and offers no cure except through the very activity – analysis – which reveals this news." (Geoffrey H. Hartman, "The Interpreter's Freud," in Easy Pieces (New York: Columbia University Press, 1985), pp. 142-3).

62 Žižek, On Belief, p. 84.

63 Žižek, On Belief, p. 148.

64 Žižek, On Belief, pp. 148-149.

65 Žižek, The Puppet, p. 22.

effects. It is thanks to the act performed by Saint Paul that Christianity as a universal religion could emerge. Paul the outsider (i.e., not part of Jesus's inner circle), Paul the betrayer indifferent to the historical Jesus and his idiosyncrasies,[66] displaces Christianity from its original context and allows it to fulfil its potential in another historical moment. For Žižek, Saint Paul is the type of a true radical whose act does to Jesus what Lenin did to Marx and Lacan to Freud. In one of his most daring (perverse?) comparisons, Žižek teases: "Paul as a Leninist: was not Paul, like Lenin, the great 'institutionaliser,' and, as such, reviled by the partisans of 'original' Marxism-Christianity?"[67]

What are we to make of Vattimo and Žižek – those two outsiders to the circle of the theologically initiated – and of their quite scandalising versions of Christianity? Once we recover from the shock of the baptised Lenin in Žižek and of the rehabilitated half-believers in Vattimo, we can reflect on the significance of their claims. We can dismiss them as unqualified to write about Christ and Christianity, as postmodern impostors invading territory they know little about. Or we can applaud them for demonstrating that it is possible to see Christianity otherwise than as a stronghold of logocentrism. Echoing Merold Westphal's words about "the Gang of Five" (a group including Nietzsche, Heidegger, Derrida, Foucault and Rorty), we can argue that Vattimo and Žižek "challenge ... religious traditions and institutions that see nothing [in them] but nasty nihilism ..., forgetting that Balaam's ass once spoke God's words."[68] I think, however, that these arguments are misguided: Christianity is no longer a logocentric fossil (*vide* Graham Ward's theology) and postmodern philosophers are not asses-turned-prophets, as Westphal's biblical allusion would suggest.[69] The significance of Vattimo's and Žižek's writing seems to lie in the fact that they try to think Christianity and culture/philosophy together, i.e., not as two distinct spheres acknowledging the importance of one for the other and interacting within various Niebuhrian types, but as two facets of the same thing, one conditioning the understanding of the other, neither remaining beyond the other, both belonging to the same moment in history. In Vattimo and Žižek, while Christianity and culture

---

66 Žižek, The Puppet, p. 10.
67 Žižek, The Puppet, p. 9.
68 Merold Westphal, "Appropriating Postmodernism," in Postmodern Philosophy and Christian Thought, ed. Merold Westphal (Bloomington and Indianapolis: Indiana University Press, 1999), p. 2.
69 Westphal alludes to the biblical account of a mysterious event involving the prophet Balaam and his ass (Numbers 22: 21-35) Balaam was a seer asked by Balak, King of Moab, to curse Israelites, which would help Balak defeat them in battle. Balaam decided to go to Moab but announced he would repeat God's words only. On the way to Moab, Balaam's ass showed inexplicable alarm, swerving from the road, crushing Balaam's leg against the wall and finally, collapsing under her owner. As Balaam – unaware that the ass saw the angel - was beating the animal, "the LORD opened the mouth of the ass, and she said unto Balaam, What have I done unto thee, that thou hast smitten me these three times?"

maintain minimal distance from each other – not unlike the minimal space of non-coincidence in the and-structure of Ward's title – they make each other possible, permit each other to occur inside time.

I use the words "permission" and "making possible" in the sense described by Michael de Certeau. In "How Is Christianity Thinkable?" of 1977, de Certeau claims that the inaugural event of Christianity – Jesus Christ – is the event as permission. In the inaugural event, Jesus dies, effaces himself and makes room for, i.e., *permits*, various historical interpretations to appear. The Gospels, as first interpretations written after the disappearance of Jesus, do not map "a 'proof' nor the 'truth' but only an effect ... . The writings ... express not the event itself but that which the event made possible in the first believers."[70] The absent inaugural event permits or initiates various responses whose fidelity is measured neither by preservation nor by repetition. None of the diverse responses to the event is identical with the event itself; nevertheless, every response acts as a testimony to the event which made the variety of responses possible. The Christian event can be located nowhere in particular but is transmitted only through "interrelations constituted by the network of expressions which would not exist without it."[71] As de Certeau emphasises, the Christ event permits or makes room for heterogeneity – for various, non-conformist interpretations within which the inaugural death and the death of past responses may be felt by uniformity-lovers as a lamentable loss. The event as permission bids Christians to recognise their limits, make room for others, confess the "necessity of others" and acknowledge "what they do not have, what they miss." To avow that imperfection stems from the permission: "[t]o permit means to disappear, and at last to die,"[72] says de Certeau. Permission is inseparable for de Certeau from "a risk of doing,"[73] i.e., from a decision to make room for that which "cannot be a priori specified," for "the unforeseeable."[74] De Certeau writes of a Christianity based on permission that it is "still capable of opening a new space; it can make possible a change in the practice of speaking and in the relation of a speaker to others."[75]

There is a striking parallel between de Certeau's Christian inaugural event understood as precious absence and Vattimo's and Žižek's interpretations of the incarnation, on which their philosophical reflections are built. When Vattimo claims that "modernity is guided by Christian scheme, i.e., weakening,"[76] he does not assume that there is a core Christian doctrine repeatedly rediscovered in its

---

70 Michael de Certeau, "How Is Christianity Thinkable?" in The Postmodern God, ed. Graham Ward (Oxford: Blackwell, 1997), p. 144.
71 de Certeau, "How Is Christianity Thinkable?" p. 146.
72 de Certeau, "How Is Christianity Thinkable?" p. 150.
73 de Certeau, "How Is Christianity Thinkable?" p. 151.
74 de Certeau, "How Is Christianity Thinkable?" p. 154.
75 de Certeau, "How Is Christianity Thinkable?" p. 147.
76 Vattimo, Belief, p. 44.

unchanged form. On the contrary, he maintains that "the substance of the Christian announcement is not Christ's revelation of an eternal truth but rather an actual historical event."[77] Significantly, the term "historical" is used "not in the sense that it is a 'real' phenomenon, but rather insofar as ... it is constitutive of our existence."[78] Or, to put it in terms of de Certeau's discourse, it is historical insofar as it permits various interpretations and enables different relationships to the world. The truth of Christianity lies in interrelations defined as kenotic. But, Vattimo hastens to add, this does not mean that weakening is "a definite undertaking" since "it may be that there is another meaning of the close relationship between faith and work of which the Bible often speaks: it is only within the history of salvation ... that the sense of the evangelical message itself is clarified and acquires meanings ...."[79] Hence, though the kenotic interpretation seems to be most reasonable in the context of the here and now, it is proposed by Vattimo together with a permission for something unforeseeable to remain in tow.

When Žižek maintains that "Christianity is, from its inception, THE religion of modernity,"[80] he does not imply that Christianity *bestows* liberty on humans. The Christ event does not invite us to project ourselves on Christ as on someone accomplishing things for us, but "opens the way" for redemption and permits our freedom to occur.

> [B]y his death, people are not directly redeemed, but given the POSSIBILITY of redemption ... . Christ does NOT do our work for us, he does not pay our debt, he "merely" GIVES US A CHANCE - with his death, he asserts OUR freedom and responsibility, i.e. he "merely" opens up the possibility, for us, to redeem ourselves ... by way of choosing to "live in Christ" - in *imitatio Christi*.[81]

In the *imitatio Christi*, we do not return to the historical Jesus, to the living person spoken about in the Gospels, but to what takes place in Christ: to the resurfaced imperfection and impotence, to something missing – some precious absence involved in time in its own, characteristic mode. In the Christ event, we are faced with a "short circuiting"[82] of temporality (i.e., the incarnation) and eternity (i.e., God the Father's traumatic mystery). As a temporal event of incarnation, Christ inevitably fails to thoroughly symbolise the trauma of God the Father, yet makes this impotence (or absence, failure) the way for the eternal trauma to interact with time. Through Christ, the trauma is not conceived as the intractable mystery hidden behind reality or outside time but as the intervention of eternity into time, an

77 Vattimo, After Christianity, p. 109.
78 Vattimo, After Christianity, p. 112.
79 Vattimo, Belief, p. 59.
80 Žižek, On Belief, p. 150.
81 Žižek, On Belief, p. 105.
82 Žižek, On Belief, p. 112.

eruption which sustains time. The Christ event is not a completed, past occurrence but a cut which disturbs closure and permits time to flow. In the Christ event, time does not circulate, rehearsing some eternal, foreseeable rhythm; rather, short-circuited by eternity, time continues with its own incompletion. If the act as the *imitatio Christi* allows us to die to the symbolic order and constitutes a new beginning, it functions like de Certeau's "risk of doing" and permission. It is in this context that we should read the following statement of Žižek about "the ultimate heroic gesture awaiting Christianity": "in order to save its treasure, it has to sacrifice itself – like Christ, who had to die so that Christianity could emerge."[83] The act as permission leads to the unforeseeable that cannot be predicted because it changes the currently available horizon of understanding.

There is a temptation to slightly modify de Certeau's phrasing as applied to Vattimo and Žižek and replace "Christianity as permission" with "permissive Christianity." Both philosophers take this temptation into account. Vattimo reiterates that his interpretation of the kenosis does not produce "an easy Christianity, but rather a friendly one, just as Christ himself preached it to us."[84] We are no longer slaves to fundamental principles nor soldiers admitted to the Church army only if we are entirely resolute.[85] We are friends of God, and friendship is never an easy thing. Žižek desperately insists that Christ's death does not have a permissive character epitomised in a message such as "now I take it for you, you can screw it up again ... now [you] can go watch hardcore movies because [you] are redeemed each time."[86] Žižek is at pains to show that his subtitle ("The Perverse Core of Christianity") does not legitimise belief in the liberating power of perversion, and that his Christianity is not a perversely permissive remake of Augustine "*ama et fac quod vis.*"

Admittedly, the Vattimo and Žižek who disclaim permissiveness coexist with the Vattimo and Žižek who make permissiveness work inside their interpretations. In other words, while their Christianity is not identical with (permissive) culture, it can transpire only through its signs, never violating, however, the minimal difference between one and the other. For Vattimo, "(the providential, perhaps) pervasiveness of pornography"[87] helps to deprive sex of its sacred, i.e., violent, aura. Weakened by ubiquitous pornography, sexuality loses its status of the "sanctuary for consciousness"[88] bestowed on it by psychoanalysis. Thus, pornography conceived as one of many signs of secularisation operates *within* the kenosis. In other words, some looseness, some dilution of morality are

---

83 Žižek, The Puppet, p. 171.
84 Vattimo, Belief, p. 55.
85 Vattimo, Belief, p. 56.
86 Slavoj Žižek, "On Divine Self-Limitation and Revolutionary Love. An Interview with Slavoj Žižek," Journal of Philosophy and Scripture 1/no. 2 (2004), p. 36.
87 Vattimo, Belief, p. 57.
88 Vattimo, Belief, p. 57.

permitted inside the Christian scheme. For Žižek, since today "transgression itself IS the norm,"[89] permissiveness becomes "part of the game"[90] and has to be taken into account. We can say that Žižek – not unlike Vattimo – gives permissiveness permission, but we can claim so only provided we understand permission as disappearance, dying or making room for something else. As Žižek contends, among the proliferation of pleasure-through-transgression gadgets available today, permissiveness is emptied of its meaning because there is virtually no limit left beyond which one could go to indulge in something forbidden or disapproved of. The way out of the disenchanted permissive world is not a new transgressive experience, but a return to the everyday, the familiar and orthodox. "What if," asks Žižek, "in our postmodern world of ordained transgression, in which the martial commitment is perceived as ridiculously out of date, those who cling to it are the true subversives? What if, today, straight marriage is 'the most dark and daring of all transgressions'?"[91] If transgression is reformulated in such a way, the beyond turns out to lie not outside the scope of the accessible but within reach, in places no one would ever call transcendental.

It seems that the idea of the (so to say) here-and-now beyond sheds some light on the reluctance displayed by Vattimo and Žižek toward the contemporary turn to religion visible in popular culture as well as in departments of philosophy and literary studies. Both Vattimo and Žižek openly denounce today's God-as-the-radically-Other craze and remain sceptical about the widespread predilection for mysticism, negative theology, and for a certain understanding of Jewish spirituality. For Žižek, the Levinasian-Deridean Otherness and the understanding of God based on it are "false" in that such Otherness, separated from men by a gap, "lead to the boring, monotonous sameness of the Otherness itself."[92] As Žižek observes, Derridean appropriation of Levinas's criticism of onto-theology paves the way for the "melancholic postsecular thought" which is concomitant with "undeconstructible form of spirituality," growing out of the matrix-like "(back)*ground*" of undecidability.[93] Far from rejecting the whole of Derrida's (or Levinas's) legacy,[94] Žižek indicates that the popularity of the Judaism-inspired

89 Žižek, On Belief, p. 41.
90 Žižek, On Belief, p. 31.
91 Žižek, The Puppet, p. 36.
92 Žižek, The Puppet, p. 24.
93 Slavoj Žižek, "Melancholy and the Act," Critical Inquiry 26/no. 4 (2000), pp. 663-4, 666; my emphasis.
94 Žižek does not totally dismiss Derrida. He calls upon Derrida's "passive decision" (from "Adieu to Emmanuel Levinas") to support his own argument about the act as a decision made irrespective of the usual predicates of metaphysical subjectivity, such as autonomy, consciousness, sovereignty (Žižek, On Belief, p. 148). While Žižek is appreciative of Derrida's differance period (he announces "Here I am tempted to suggest a return to the earlier Derrida of differance." [Žižek, The Puppet, p. 141.]), he is very critical of Derrida's "postsecular messianism." Against Derrida's future perennially beyond our reach (a messianic future that is always to come), Žižek proposes his temporality based on Pauline "already, but not yet," in which "the big thing happened" (Žižek,

openness to the Other and its unconditional call is the result of belittling or effacing of the most precious aspect of Jewish spirituality: its "unique collective experience."[95] Vattimo is even more adamant about "a sort of predominance of Judaic religiosity in the return to religion into contemporary thought," whose effect lies in the fact that "the total otherness of God appears to be affirmed at the expense of any recognition of novelty in the Christian event."[96] Though Vattimo appreciates Levinas's attempts to philosophise outside metaphysics, he disapproves of Levinas's ideas on God, claiming that he erases the differences between historical times when he makes every historical moment related immediately to eternity/the Other in a purely vertical dimension. Likewise, Vattimo criticises the leap-of-faith religion since it discovers God in crisis or defeat of understanding, as alterity separated from men by an "unsurpassable limits"[97] and accessible only through a leap beyond reason. Vattimo's criticism of the leap of faith runs parallel to Žižek's deep scepticism of ecstatism, mysticism or Gnosticism, which share the belief in a lost perfection, in the unfortunate Fall and separation from the divine or in the divine secret inaccessible to the uninitiated. By and large, Vattimo and Žižek – in a way heretical thinkers themselves – criticise the contemporary focus on the heterodox (the heretical, the iconoclast), in which the sphere of the beyond lies outside the realm of the human, and concentrate on the mainstream – not to say mundane – spirituality for which the beyond resides in the commonplace, in the detail, on the surface. "God resides in details," says Žižek, "in the overall drabness and indifference of the universe, we discern the divine dimension in barely perceptible details – a kind smile here, an unexpected helpful gesture there."[98] Vattimo entitles a chapter of his *After Christianity* "God the Ornament," by which he implies that divine existence – like the one described in the Book of Revelations – dissolves into secondary qualities to which our senses are well-accustomed: the ornamental, the volatile, the fleeting.

In other words, Vattimo and Žižek bring back to the surface something that was always – in one way or another – near the surface of the Western culture. They concentrate on the religion whose status in the contemporary world is comparable to the status of a long preface: it is perceived but treated as unnecessary paraphernalia no sane (average) person (reader) bothers about. They do to Christianity what Gerard Genette did to paratexts[99]: put in the spotlight something

---

"On Divine Self-Limitation," p. 36) – the Messiah was here – but not everything has been decided.

95 Žižek, The Puppet, p. 8.

96 Vattimo hastens to emphasise that his opinion "has no anti-semitic intention whatsoever." (Vattimo, Belief, p. 84)

97 Vattimo, Belief, p. 25.

98 Žižek, On Belief, p. 98.

99 Gerard Genette's "paratext," discussed in the book Paratexts: Thresholds of Interpretation, names "a heterogeneous" and constantly changing group of practices and discourses, located in the "'undefined zone'" of titles, subtitles, running titles, notes, chapter titles, prefaces, author names, blurbs, book covers, dust jackets, etc. Paratexts seem to share certain features with Vattimian and

that – though noticed or even analysed before – remained a taken-for-granted, perceived but not exhaustively explored, fringe phenomenon. While Genette changed the status of titles, prefaces etc. from something that remains tacitly there into a troublesome, thought-provoking field of studies that cannot be easily bypassed, Vattimo and Žižek accomplish a similar thing with Christianity. They put their version of encultured Christ in a position which in my discussion of Graham Ward I described as the precarious space of titles – the confusing, vaguely demarcated *front*-line of a text. Like titles, which according to Genette are addressed "to many more people than does the text, people who in one way or another receive and transmit it, and thereby contribute to its circulation,"[100] their paratextualised Christianity figures as the detail of the Western culture through which the rest of the culture can be opened, accessed and changed. There is always the risk, however, that the paratextualised Christianity will be a mere flicker, a

---

Žižkean Christianity. First, paratexts defy categorisation into the inside and the outside; they function within a wide range of degrees of intimacy with the text and embrace "peritext" (remaining on and between covers, close around the text or inserted into its interstices) and "epitext" (maintaining distance to the text, i.e., located at the outside of a bound volume, in the form of interviews, conversations with the author, reviews, etc.); Gerard Genette, Paratexts: Thresholds of Interpretations, trans. Jane E. Levin (Cambridge: Cambridge University Press, 1997), p. 5. In a way, paratexts problematise the easy distinction between the beyond and the within. Second, the function of paratexts is to ensure that the text they accompany is present in the world, i.e., to reinforce, advertise and "adorn" the text (Genette, Paratexts, pp. 1-3). To produce the desired effect, however, a paratext should remain subtle if not weak; over-imposing, too strong, overly controlling paratexts become impediments, as was the case with certain too-clever titles which deterred Genette from reading the books they named (Genette, Paratexts, p. 94). Like Vattimo's and Žižek's Christianity, a paratext is marred when played too strong and effective when given only a "light touch" (Genette, Paratexts, p. 410). Third, paratext can exist otherwise than as threshold of a substantial "something else," i.e., of something whose presence is taken for granted. Titles can and do exist without texts – their "natural" twin term (Genette, Paratexts, pp. 3-4.), which means that in paratext, our usual expectation of its characteristic "transaction" (Genette, Paratexts, p. 2.) can be frustrated because together with a paratext we can get "nothing" instead of a positively existing "something." Also, among titles which function in a normal way, i.e., those which accompany an available rather than non-existent text, there are ironic titles (a subcategory of the most popular type of thematic titles), which display "provocative absence of thematic relevance" (Gerard Genette "Structure and Function of the Title in Literature," Critical Inquiry 14/ no. 4 (1988), p. 713). In other words, ironic titles identify or name – as titles are supposed to do according to Genette – something that does not appear in the text. Simultaneously, paratexts figure as distinctly fragile and impermanent: they may "disappear, definitely or not"; they may be rapidly cut (shortened) or gradually eroded. (Genette, Paratexts, p. 6.) Paratexts "do not constitute a uniformly unvarying and systematic presence," (Genette, Paratexts, p. 3) and may become absent. Seen in that way, paratext stands for some kind of stubborn persistence of that which names absence and reveals impotence. Impotence or imperfection is inscribed into Genette's study because he leaves out three major fields of paratextual elements (translation, serial publication and illustration), making his study "incomplete" (Genette, Paratexts, pp. 404-6). The paratextual weakness, its potential for disappearance and its openly declared imperfection are also features of Vattimo's and Žižek's Christianity.

100 Genette "Structure and Function," p. 707.

grimace, a fleeting presence on the philosophical shelf. Will it not resemble books described by Italo Calvino, books "You Needn't Read, Books Made For Purposes Other Than Reading, Books Read Even Before You Open Them Since They Belong To the Category Of Books Read Before Being Written [...] Books That Everybody's Read So It's As If You Had Read Them, Too"?[101] And if so, will Vattimo's and Žižek's encultured Christ be permitted into the readers' minds supposing that Calvino's belligerent, strong and pushy books were not? In Calvino, the books are left behind by the reader for the sake of another, new story. But it is precisely the risk of disappearance – the risk *taken* by Vattimo and Žižek – that persuades us to take their ideas seriously.[102]

---

101 Italo Calvino, If on a Winter Night a Traveller (Orlando, Florida: Harvest Books, 1982), p. 5.
102 This idea is readable in the context of one of Vattimo's remarks (that only when we abandon the idea of ultimate foundation and strong structures, can we recover Christianity and take it seriously, [Vattimo, After Christianity, pp. 5, 7]) and in the light of one of Žižek's thoughts (that we love something (someone) "BECAUSE of his limitation, helplessness ... ." [Žižek, On Belief, p. 147]).

Marta Zając

Recyclable Adam? On Dustbins of History and "the Dust of the Ground": Jean Baudrillard's and Thomas Merton's Notions of Tradition

In one of her essays Susan Sontag blames contemporary art for what she calls "[a] coquettish, even cheerful nihilism," a mental state in which "[d]iscovering that one has nothing to say, one seeks a way to say that."[1] To see Jean Baudrillard's philosophical output, in paricular his book *The Illusion of the End*, as a part of the trend Sontag mentions would be for one or another reason unjust. However, the ease with which postmodern thinkers, Baudrillard in a spearhead, juggle with the ideas of fundamental or at least, let us be honest, *some* significance – is disquieting. Concepts like reality, identity, God are given so little weight that their presumed end cannot actually make a considerable impact. As a result, the only viable form of engagement left to cultural theory writers is the invention of the new rhetoric, which would add an air of freshness to all too familiar catastrophes of the postmodern age. The rhetoric Baudrillard employs in *The Illusion of the End* has its origin in the site strikingly peripheral to the apparently inextinguishable hopes and ambitions of the human spirit, that is in the refuse dumps. In Baudrillard's discourse the litter is not any longer relegated to discreetly located containers, since now the litter itself contains the attributes that the postmodern man is invited to recognise as his own.

First of all, Baudrillard defines history as a *dustbin*.[2] From the vantage point of postmodernity history contains nothing but *waste*: "defunct ideologies, bygone utopias, dead concepts and fossilised ideas ... the great empires, the grand narratives, the great systems made obsolete by their own gigantism ...."[3] Consequently, postmodernity is experienced as a permanent emergency call: recycle or perish. "[W]e can either perish under the weight of the non-degradable waste ... or else recycle,"[4] Baudrillard's commentary reads. As ecology and industry teach us, there are two solutions to the waste problem: recycling and the incinerator. Both methods are recognised by Baudrillard as postmodern ways of dealing with the problem of intellectual refuse.[5] At the same time the presumed solution to a problem causes another problem, not less serious. Due to the presence

1 Susan Sontag, "Aesthetics of Silence," in Styles of Radical Will (New York: Dell Publishing, 1969), p. 12.
2 Jean Baudrillard, The Illusion of the End, trans. Chris Turner (Stanford, California: Stanford University Press, 1994), p. 26.
3 Baudrillard, The Illusion of the End, pp. 26-27.
4 Baudrillard, The Illusion of the End, p. 27.
5 Baudrillard, The Illusion of the End, p. 27.

of the recycling practice "*History will not come to an end*," [6] Baudrillard prophesies. Nothing disappears ("Nothing one thought superseded by history has really disappeared"[7]), which brings about a kind of "environmental catastrophe" in the universe of ideas: the "pollution" of mental space, its saturation.

Nothing old disappears: nothing new can be crystallised in the "polluted" air of postmodernity. Due to the "hyperdensity of cities, commodities, messages and circuits"[8] the circulation of ideas is not possible. Past ideas, unless recycled, bring about the "saturation of exchanges"[9] and the consequent paralysis of thought. Accordingly, in a striking contrast to the position urban communities of the West aspire to, Baudrillard writes about their inertia and silence. To be exact, silence falling on the cities, which to the evidence of the senses are humming with activity and life, is the silence of the *meaningless*: "all the atoms of meaning get lost in space. Each atom pursues its own trajectory to infinity and is lost in space."[10]

While following Baudrillard's presentation of the developed societies of the West, one cannot escape the impression that his analysis of the *post-modern* condition implies a kind of *post-mortem* perspective. He inspects "inert matter,"[11] the *dead* body of the social. In many ways postmodern man assimilated the order of the lower-rank reality, the order of objects, the order of recyclable waste. Apparently, the actual message of Baudrillard's description does not concern the drama of the prospective situation when history (or whatever still remains of it on the intellectual orbit of the West) *will not come to an end.* The world Baudrillard examines *has already come to an end* by the very way it seemingly continues to move on. At a certain point in history man "choked" (to death) on the excess of information. Now, it is the mere matter of the past ideas that he will endlessly (and mechanically) reiterate. Their spirit is long gone.

Two related questions that Baudrillard's vision ultimately provokes read: 1) why is it postmodernity that defines itself through the recycling practice, and 2) why is it postmodernity that postulates the very problem of recycling *ideas*? On the one hand, the questions fall into the scope of broadly-discussed and actually well-developed problematics of the effects the spread of technology has on our functioning in the world. The impact of technology on man's psyche is not only a theoretical issue in culture-oriented volumes but finds place in practical psychology and psychiatry. Antoni Kępiński, a leading Polish psychiatrist, mentions, for instance, changes which occur in language under the pressure of the ubiquitous presence of technology: "Our view of the world depends on the manner we act upon it. ... A shift from natural environment to technological *milieu* creates a new

6 Baudrillard, The Illusion of the End, p. 27.
7 Baudrillard, The Illusion of the End, p. 26.
8 Baudrillard, The Illusion of the End, p. 3.
9 Baudrillard, The Illusion of the End, p. 3.
10 Baudrillard, The Illusion of the End, p. 2.
11 Baudrillard, The Illusion of the End, p. 3.

perception of man as an element in the technology-oriented world. Among others, the phenomenon gets reflected in everyday language ... when the verbs commonly applied to objects find their usage in the expressions related to people, e.g. 'remove,' 'put off,' 'take aside.'"[12] At the same time, however, another problem looms large. Recycling, generally, concerns the leftovers, and the recycling of culture, specifically, refers to what remains when ideas, theories, systems, or even values, have been "used up." One should consider at this point the following question: how to make sense of the idea of "culture recycled," if at all? How do we "consume," "apply" and finally "use up" mental products? I would like to approach these problems taking into account the concept of "education." In my view, education-related issues though absent in Baudrilard's argument are still relevant to some of his claims. For example, education as knowledge *acquisition* sheds new light on the definition of intellectual waste. Certainly, you may dispose only of what already somewhat belongs to you. On what grounds can we claim ideas our own?

The linear order of time creates the illusion of the accumulation of goods. Thus, we tend to believe that the wisdom[13] of the past ages is ours by the very force of our temporal position in a succession of historical periods. However, one does not become a successor to past ideas as a matter of simple (intellectual) routine. The lineage is a product of the physical union of bodies rather than spiritual communication between the ages. Søren Kierkegaard, in Fear and Trembling, makes a helpful remark on false preconceptions about educational process,

> Suppose someone wanting to learn to dance said: "For hundreds of years now one generation after another has been learning dance steps, it's high time I took advantage of this and began straight off with a set of quadrills." One would surely laugh a little at him; but in the world of spirit such an attitude is considered utterly plausible. What then is education? I had thought it is the curriculum the individual ran through in order to catch up with himself; and anyone who does not want to go through this curriculum will be little helped by being born into the most enlightened age.[14]

As early as at the very beginning of the quoted sentence we read about "someone" and thus find a shift towards individuality. Typically for the Christian interest in

12 "Widzimy otaczający świat tak, jak na niego działamy. ... Zmiana środowiska naturalnego w techniczne powoduje, że także na ludzi zaczyna się patrzeć jak na części składowe świata technicznego. Potoczny język ... oddaje to zjawisko w dość powszechnym dziś zastosowaniu wobec ludzi wyrażeń odnoszących się do przedmiotów martwych, np. 'wysunąć,' 'przesunąć' ... 'zdjąć,' 'ustawić,' 'postawić' itp." Antoni Kępiński, Rytm życia (Warszawa: Sagittarius, 1993), p. 126; translation mine.

13 Or, alternately, the stupidity, as Baudrillard also suggests: "Who will rid us of the sedimentation of the centuries of stupidity?" Baudrillard, The Illusion of the End, p. 26.

14 Søren Kierkegaard, Fear and Trembling, trans. Alastair Hannay (Harmondsworth, New York: Penguin Books, 1985), p.75.

persona (whose boundaries should be respected and never erased by the pressure of the outside world), Kierkegaard persistently takes as his hero an individual, be it a concrete figure, like Abraham or Isaac, or a type, like Kierkegaard's famous "knight of faith." In this respect Kierkegaard's perspective complements Baudrillard's picture of the world. Needless to say, society depicted as "silent mass of immanence" can hardly make for a living community, a group of individuals.

Obviously, the mentioned focus upon individual is not a matter of a single word, but first of all is present in Kierkegaard's understanding of education as a kind of "path towards oneself." The end of the educational process is for him self-knowledge, learning about oneself. To be exact, Kierkegaard writes about "catching up with oneself." That may imply the inner drama of uncertainty, a sense of doubt about the very truth of one's own existence, the highly subjective feeling that I am, but I am always one step ahead or behind myself, within hand's reach but still somewhat beyond grasp. Perhaps what is needed to overcome this tension (and ultimately "catch up with oneself") is a certain logic and continuity of human development. Kierkegaard rightly notices, "lower natures forget themselves and become something new. Thus the butterfly has altogether forgotten that it was a caterpillar, perhaps it can so completely forget that it was a butterfly that it can become a fish."[15] It seems that in Baudrillard's model of reality a butterfly can easily become a fish, the main reason for which is the provisional and non-obligatory nature of identity as such. Neither butterfly nor fish are true, after all. The curse of postmodernity is not the very recognition of *hyperreality* (whose presence can hardly be denied) but the very way the phenomenon is received. Resigned acceptance or dull enthusiasm frequently happen to be the only commentary. The search for the criteria to distinguish between the authentic and artificial, even at the micro-level, that is to meet the current needs of a sane individual, is dismissed as on the one hand too demanding and on the other unbecoming to the mind which has just discovered that – *simulacrum is true.*

Psychiatrists reflect on and diagnose that state in a variety of ways. Andrzej Leder, a psychiatrist and a columnist in *Res Publica Nowa*, evokes a well-known image of media- and technology-dominated world where mind is flooded with purely accidental and unrelated images. Their flow, from behind the anonymous TV or computer screen, marks the collapse of personal security and integrity structures. The idea to consider is that media-born narratives, though discreetly deposited in the dark corners of the mind, make little sense for an individual. These visions hardly enter the pattern of a personal, unique, inimitable life history and thus contribute to the inner loss and disorientation ("Confessions of people who are not a part of my life, photos of the dishes I will never make, eruptions of volcanoes about which I know nothing ... the world composed of confused visions, behind

---

15 Kierkegaard, Fear and Trembling, p.72.

which there remain a host of conflicting spiritual projects."[16]) In Kępiński's explication of the conditions for successful self-formation one finds a related issue. The necessary stage in the mentioned process is "knowing one's way about." The mechanism Kępiński describes is valid on all levels of being and its organisation from the most elementary (e.g. cells) up to the most complex (e.g. humans): "Informational metabolism is the precondition for other forms of contact with the outside world like assimilation and reproduction. Before the world becomes 'mine' either in the sense of building up the substance of the body or through the sexual union, it has to become 'familiar.' An organism needs to know how to move on."[17]

The insights of psychiatry help one combine Baudrillard's image of postmodern condition with Kierkegaard's concept of education. When the human mind loses its natural abilities to negotiate the borders with the outside world, when the apparently outer conditions of cultural space: hyperdensity, saturation, dissemination, govern the composition of mental space, the educational process in the sense promoted by Kierkegaard, that is as a path towards oneself – is blocked. The saturation of mental space renders impossible both the circulation of ideas and the constitution of one's self. For Kępiński this particular state, diagnosed as a neurotic symptom, calls for a therapy or other forms of treatment.[18] In Baudrillard one will search for remedies in vain. As mentioned, a way out *The Illusion of the End* suggests is the construction of the "incinerator": an example of purely imaginary solutions given to painfully concrete social dilemmas. Once again, one may wonder about Baudrillard's insistence upon man's fall into the sphere prepared for the scraps of the cultural processes that man's own hand and mind put into motion.

It may be revealing to notice how similar problems, in particular the question about the capacity of cultural space for preserving old ideas, are dealt with in other discourses. On the assumption that the postmodern perspective, however central to contemporary discourse it remains, still leaves some space for other beliefs and propositions, I wish to broaden my discussion with the views of Thomas Merton.[19] An American poet and writer, but in the first place a Trappist,

---

16 "Wyznania kogoś, kogo w moim życiu tak naprawdę nie było, obrazy potraw, których nigdy nie zrobię, wybuchy wulkanów, o których nic nie wiem ... świat składający się z sąsiadujących ze sobą, przemieszanych rojeń ..., za którymi kryją się niewiadome ilości zupełnie różnych i w żaden sposób do siebie nie przystających projektów duchowych." Andrzej Leder, "Na początku był chaos," Res Publica Nowa 86/ no. 11 (1995), pp. 29-31; translation mine.

17 "Metabolizm informacyjny jest wstępnym krokiem przed wejściem w kontakt asymilacyjny i reprodukcyjny ze światem otaczającym. Nim stanie się 'moim' w sensie tworzenia substancji własnego ustroju lub seksualnego złączenia, musi stać się 'moim' w sensie możności orientowania się w nim. Ustrój musi wiedzieć, jak się w nim poruszać." Kępiński, Rytm życia, p. 32; translation mine.

18 See Kępiński, Rytm życia, p. 127.

19 One explanatory remark should be added at that moment. Building up a contrast between Baudrillard, a postmodernist, and Merton, a Catholic monk and contemplator, is justified. In his reflection on the developed societies of the West, Baudrillard leaves no doubts that God is not there

Merton both in his life and work reveals the drama of man balancing between the ideals of contemplation and the inner urge to confront the tensions of the world apparently left behind the gates of the monastery.[20]

In Merton the conflict between progressive and conservative tendencies, which evidently underpins the waste problem Baudrillard outlines, is resolved through the concept of Christian tradition as simultaneously form of revolution: "The biggest paradox about the Church is that she is at the same time essentially traditional and essentially revolutionary ... Christian tradition, supernatural in its source, is something absolutely opposed to human traditionalism. The living Tradition of Catholicism is like the breath of a physical body. It renews life by repelling stagnation. It is a constant, quiet, peaceful revolution against death."[21] Without doubt dogmas remain the very foundation of the earthly dimension of the Church; at the same time Merton compares the presence of Christian tradition to a flow of fresh air through the physical body it keeps from decay. In his view, the steady circulation of dogmas protects the community of believers against spiritual exhaustion and routine. Thus conceived the meaning of tradition is fully positive. The same presence of tradition which is said to ruin social structures keeps alive the body of the Church. Obviously it is not yet the presence of the *same* tradition that Merton and Baudrillard write about. In Baudrillard tradition is defined along the temporal axis as a relation between past and present. Merton's concept of tradition in turn assumes a-temporal dimension of eternity and thus centres on the co-presence of the divine and human element. For Merton dogmas, the inscription of Christian tradition, are a sign of God's Living Presence. Simultaneously, individual moves and choices remain of utmost significance. On the one hand, the human element is subjected to all the changes necessary for maintaining the divine essence of the Church. On the other, the continuing presence of dogmas demands individual believers' genuine effort to incorporate Christian truths and to make those truths the living experience.

(see, for example, Baudrillard, The Illusion of the End, p. 92.). Maintaining the opposition between postmodernist philosophy and Christian theology at large seems less evident. It is not a hopeless enterprise to negotiate the distance between one's belief in the Living God of the Catholic Church (with all the dogmas maintained) and some of the strictly postmodern intuitions. In particular, I have in mind Tomáš Halik's recent work, published in 2002, Co je bez chveni, neni pevné. Labyrintem sveta s virou a pochybnosti [What Is not Unstable, Will not Last. Through the Labyrinth of the World with Faith and Doubts]. Halik, a Catholic priest, philosopher and acknowledged theologian, is also a keen reader of Kant, Wittgenstein, Freud and Nietzsche.

20 Even a cursory glance at Merton's body of work is enough to notice the variety of planes and problems there included. One will find a broad scope of themes, ranging from strictly spiritual issues (e.g. Th. Merton, Seeds of Contemplation), up to current problems of monastic discipline as well as feminism or media analysis (e.g, Th. Merton, Sources of Contemplation, Th. Merton, Cz. Miłosz, Letters).

21 Merton, New Seeds of Contemplation (the Abbey of Gethsemani: New Directions, 1961), pp. 142-143.

When the problem of individualism is focused on what should be explicated is the manner Merton's belief in the value of an individual life and efforts agrees with the ideals of self-renunciation he simultaneously promotes. For Merton the ideal state for contemplation demands cutting off all the ties with the "world" conceived by him as the place aiming at the production of "false selves." In Merton's understanding a "false self" is "the man that I want myself to be but who cannot exist, because God does not know anything about him ...."[22] At its extreme Merton's reflection reaches a truly imposing vision of humanity whose very subsistence depends on the efforts of a few. If attained, the state of personal freedom from artificial (unknown to all-knowing God) identities functions as a kind of pivot on which the whole world rests: "I wonder if there are twenty man alive in the world now who see things as they really are. That would mean that there were twenty men who were free, who were not dominated or even influenced by any attachment to any created thing or to their own selves or to any gift of God ... I don't believe that there are twenty such men alive in the world. But there must be one or two. They are the ones who are holding everything together and keeping the universe from falling apart."[23]

However unfamiliar one may feel with these concepts (after all, the fruits of purely contemplative life), at some places Merton's visionary ideas bear a striking resemblance to Baudrillard's intuitions about postmodernity. Regardless of the scale, level or mode of being considered, there needs to be a certain force for things to subsist (and make sense). The concept of "gravitation" Baudrillard employs seems to be his equivalent of (missing) God. "Gravitation" is necessary for things to mean and to last: "we ... passed beyond a certain horizon in which the real is still possible because gravitation is still strong enough for things to be reflected and thus in some way to endure and have some consequence ... beyond this gravitational effect, which keeps bodies in orbit, all the atoms of meaning get lost in space."[24] Typically for a Christian, Merton sustains his belief in the weight of personal freedom from simulacra. Typically for a postmodernist, Baudrillard allows for the idea of mass liberation from any need for the real: "We are 'liberated' in every sense of the term, so liberated that we have taken a leave of a certain space-time, passed beyond a certain horizon in which the real is possible."[25] For me, it is this opposition which imposes further choices. Among others, Baudrillard's concept of the "I" as modelled on the artificial and the subsequent subjection of the "I" and its products to the mechanical process of recycling. Baudrillard's rhetoric of waste is more than a mere tool for expression. It provides a mirror for the last reflection of the atoms of meaning speeding into the space of no consequence.

---

22 Merton, New Seeds of Contemplation, p. 34.
23 Merton, New Seeds of Contemplation, p. 203.
24 Baudrillard, The Illusion of the End, pp. 1-2.
25 Baudrillard, The Illusion of the End, p. 1.

Justyna Pacukiewicz

Ruskin's Recycling of the Middle Ages

The aim of this paper is to analyse the mechanism of significance production in John Ruskin's interpretation of the Middle Ages. Ruskin's texts will be read through the prism of the semiotic understanding of culture offered by two texts of the Tartu school: *The Structure of the Artistic Text* and "On the Semiotic Mechanism of Culture." The methodological base assumed is the conception of Yuri Lotman and Boris Uspensky concerning modelling systems and the differentiation of culture as the semiotic mechanism.

Lotman distinguishes between primary and secondary modelling systems. The former is a natural language related to reality, while the latter is a language of description related to all other languages of art and, in a wider sense, languages of culture. [1] With Uspensky Lotman analyzes culture as organising the world surrounding man with a language as its centre. Culture can be characterised by its attitude to signs and signification. The authors distinguish between cultures which put stress on the plane of content and those that emphasize the plane of expression. The former are characterised by a free relation between the signifier and the signified. By contrast, a fixed relation between the plane of content and the plane of expression is present in cultures that put stress on the plane of expression.[2]

In view of these conceptions, both the Middle Ages and the Victorian period are secondary modelling systems. This means that they are defined as cultural phenomena with their products to be treated as the realisation of culture and to be interpreted as signs. In her thorough analysis entitled *A Dream of Order* Alice Chandler interested in the study of medievalism, tracing its origins back from the period of Renaissance till modern times, notes that in the Victorian period the Middle Ages were used as "a standard by which to measure and modify current-day life."[3] The idea of the Middle Ages in Ruskin's texts is definitely an example of their idealisation, and therefore can be viewed from the perspective of the medieval revival within the Victorian period as a secondary modelling system. At that time the Middle Ages reappear in Victorian poetry, Pre-Raphaelite painting and architecture. They inspire religious, social and critical thinking.

Ruskin's discourse, which can be analysed as being representative of its age, is characterised by an attempt to reach clarity. Yet the procedures of grouping,

---

1 Peeter Torop, "Cultural Semiotics and Culture," *Sign System Studies* 27 (1999), pp. 9-23: 1 Aug. 2006 < http://www.ut.ee/SOSE/7torops.html>

2 Jurij Łotman and Borys Uspieński, "O semiotycznym mechanizmie kultury," in M. R. Mayenowa and E. M. Janus, eds., *Semiotyka kultury* (Warszawa: Państwowy Instytut Wydawniczy, 1977).

3 Alice Chandler, *A Dream of Order. The Medieval Ideal in Nineteenth-Century Literature* (London: Routledge & Kegan Paul, 1971), p. 12.

selecting and contrasting are not a matter of a mere rhetoric. The broader philosophical perspective for the interpretation of the Middle Ages appears to be based on some general ontological assumptions. They concern the view of history, religion and culture. History, including the history of art and religion, is understood as a collection of periods. Ruskin discerns "A Trinity of Ages," namely, the Classical, the Middle and the Modern Age.[4] The types of religion are also defined in chronological order: "We Europeans, then, have had three great religions: the Greek, which was the worship of the God of Wisdom and Power; the Medieval, which was the worship of the God of Judgement and Consolation; the Renaissance, which was the worship of the God of Pride and Beauty,"[5] and the fourth religion, typical of the English, the worship of "the 'Goddess of Getting-on,' or 'Britannia of the Market.'"[6]

According to Ruskin, the need for religion, no matter whether the faith is directed towards Christianity or towards "the Goddess of Getting-on" is a characteristic feature of culture. He differentiates between civilisations on the basis of their attitude towards religion: "I say that Classicism began wherever civilisation began, with Pagan Faith. Medievalism began, and continued, wherever civilisation began and continued to *confess* Christ. And, lastly, Modernism began and continues to *deny* Christ."[7]

For Ruskin the border line between Medievalism and Modernism is signified by the art of Raphael, who "as the chief representative of the Christian artists of his time, … had neither religion nor originality enough to trace the spirit of poetry and the spirit of philosophy to the inspiration of the true God, as well as that of theology; but that, on the contrary, *he elevated the creations of fancy on the one wall, to the same rank as the objects of faith upon the other.*"[8] Raphael's art becomes for Ruskin a mark of change: "… I say that a change took place, about the time of Raphael, in the spirit of Roman Catholics and Protestants both; and that change consisted in the *denial* of their religious belief, at least in the external and trivial affairs of life, and often in far more serious things."[9]

By qualifying himself and his contemporaries as "moderns,"[10] Ruskin becomes the one who stands out, with a capacity to tell right from wrong, beauty

---

4 John Ruskin, "Lecture IV. Pre-Raphaelitism," in John Ruskin, *Pre-Raphaelitism. Lectures on Architecture and Painting* (London: J. M. Dent & Sons Ltd., New York: E. P. Dutton & Co, 1907), reprint ed. Ernest Rhys (Everyman's Library. Essays and Belles Lettres. Ruskin's Pre-Raphaelitism and Other Essays upon Art: 1920), p. 153.

5 John Ruskin, "Traffic," in John Ruskin, *Selections and Essays,* ed. & intro. Frederick William Roe (New York, Chicago, Boston: Charles Scribner's Sons, 1918), p. 300.

6 Ruskin, "Traffic," p. 302.

7 Ruskin, "Lecture IV. Pre-Raphaelitism," p. 155.

8 Ruskin, "Lecture IV. Pre-Raphaelitism," p. 163.

9 Ruskin, "Lecture IV. Pre-Raphaelitism," p. 155.

10 E.g. John Ruskin quoted in: Peter Smith, *The Literary Criticism of John Ruskin*, ed. & select. Harold Bloom (Gloucester, Mass: The Anchor Books, 1969), p. 100.

from ugliness, art from non-art and explain what he and his reader have in common: "I shall only endeavour to analyse the idea which I suppose already to exist in the reader's mind."[11]

Ruskin's writing on the Middle Ages definitely bears the marks of the search for origin, so typical of evolutionary thinking. He shows man as similar to certain species, turning from a caterpillar into a moth: "That we know more than 13th-century people is perfectly true; but that is not the essential difference between us and them. We are different kind of creatures for them, - as different as moths are different from caterpillars; and different in a certain broad and vast sense, ... in a sense so great and clear that we are enabled to separate all the Christian nations and tongues of the early time from those of the latter time, and speak of them in one group as the kingdoms of the Middle Ages."[12]

In the Victorian period progress was the name applied to the changing face of socio-economic structure epitomised by the Crystal Palace exhibition. The evolutionary thought of Spencer saw the evolution of man directed towards perfection.[13] Ruskin views perfection from a different angle: "the first cause of the fall of the arts of Europe was a relentless requirement of perfection, incapable alike either of being silenced by veneration for greatness, or softened into forgiveness of simplicity."[14] It is not perfection but imperfection that has to be accepted as the truth governing life: "imperfection is in some sort essential to all that we know of life. It is the sign of life in a mortal body, that is to say, of a state of progress and change. Nothing that lives is, or can be rigidly perfect; part of it is decaying, part nascent."[15]

Ruskin sees the absurdity of perfection being taken as the aim of contemporary production. Firstly, it can never be reached, as perfection is God's attribute.[16] Secondly, it leads to the mechanisation of man, whose work becomes the work of a soulless machine. The great man is the one who accepts his imperfection. That was the case with the Medieval worker summoned to the service of Christianity, whose creations bear the marks of individuality: "to every spirit which Christianity summons to her service, her exhortation is: Do what you can, and confess frankly what you are unable to do; neither let your effort be shortened for fear of failure, nor your confession silenced for fear of shame."[17]

---

11 Ruskin, "The Nature of Gothic," p. 139.

12 Ruskin, "Lecture IV. Pre-Raphaelitism," p. 153.

13 See Robert M. Young, "Herbert Spencer and 'Inevitable' Progress," in Gordon Mardsen, ed., *Victorian Values*, (New York: Longman, 1992).

14 Ruskin, "The Nature of Gothic," p. 157.

15 Ruskin, "The Nature of Gothic," pp. 156-157.

16 "If we pretend to have reached either perfection or satisfaction, we have degraded ourselves and our work. God's work only may express that; but ours may never have that sentence written upon it, - 'And behold, it was very good.'," Ruskin, "The Nature of Gothic," p. 165.

17 Ruskin, "The Nature of Gothic," p. 146.

Medieval art is the expression of the acceptance of fallen nature of man, typical for Christianity, while modern culture is contrasted to it as the culture without God. The problem of mechanisation and aiming at perfection appears in many texts by Ruskin which are based on the analysis of medieval and contemporary products of culture. Contemporary man is presented as divided: "It is not, truly speaking the labour that is divided; but the man: – Divided into mere segments of men – broken into small fragments and crumbs of life; so that all the little piece of intelligence that is left in a man is not enough to make a pin, or a nail, but exhausts itself in making the point of a pin, or the head of a nail."[18]

The medieval sculptor is contrasted with a modern worker on the basis of his work being interpreted as a sign of "life and liberty."[19] According to Lotman, examining the works of art is about reconstructing the outlines of the model of the world.[20] Art is a modeling system and can be described as a kind of secondary language, while an artistic work is a text in this language.[21] Ruskin reconstructs the outlines of the model of the Middle Ages by analyzing gothic art. For instance, he asks us to "gaze upon the old cathedral front" and discover "ugly goblins, and formless monsters, and stern statues, anatomiless and rigid."[22] However, in this case the reader is not confronted with the actual work, but rather with a work abstracted as gothic by Ruskin. In his recycling of the Middle Ages Ruskin often speaks in generalized terms not only on art but also on the nature of the society that has created it.

The very method of formulating the title of the second volume of *The Stones of Venice* as "The Nature of Gothic" implies discerning Gothic as a phenomenon with certain characteristic features. In fact, Ruskin clearly states the reason for his being interested in gothic, which is the creation of the universal idea of gothicness, as he calls it: "And it is this Gothicness, - the character which, according as it is found more or less in a building, makes it more or less Gothic, - of which I want to define the nature; and I feel the same kind of difficulty in doing so which would be encountered by any one who undertook to explain, for instance, the nature of Redness, without any actually red thing to point to, but only orange and purple things."[23]

Ruskin introduces a normative formula, which is to instruct the reader in the correct understanding of gothic art. He brings medieval art down to six characteristic or moral elements of Gothic: "1. Savageness. 2. Changefulness. 3. Naturalism. 4. Grotesqueness. 5. Rigidity. 6. Redundance. These characters are

---

18 Ruskin, "The Nature of Gothic," pp. 150-151.
19 Ruskin, "The Nature of Gothic," p. 149.
20 Jurij Łotman, *Struktura tekstu artystycznego*, trans. Anna Tanalska (Warszawa: Państwowy Instytut Wydawniczy, 1984), p. 30.
21Łotman, *Struktura*, p. 19.
22 Ruskin, "The Nature of Gothic," p. 149.
23 Ruskin, "The Nature of Gothic," pp. 138-139.

here expressed as belonging to the building; as belonging to the builder, they would be expressed thus: -1. Savageness, or Rudeness. 2. Love of Change. 3. Love of Nature. 4. Disturbed Imagination. 5. Obstinacy. 6. Generosity. And ... the withdrawal of any one, or any two, will not at once destroy the Gothic character of a building, but the removal of a majority of them will."[24]

Gothic architecture remains a sign, which he takes as the expression of the individualism of its maker. The maker, in turn, is presented as shaped by the natural environment he lives in. He is also submerged in medieval religiosity. Ruskin's reader, put in the position of a migrating bird to see "the gradation of the zoned iris of the earth in all its material vastness,"[25] is offered a conception of a man of the North and his art as determined by nature. The idea of nature in Ruskin is not limited to denotation of the natural environment of flora and fauna but it also refers to an intrinsic spiritual dimension of man. "Christianity having recognised, in small things as well as great, the individual value of every soul"[26] is expressed in medieval art. This "individual value" is represented in medieval ornament, where human thought rather than mechanical or perfect performance is observed. The medieval and the modern arts are abstracted by Ruskin in the following distinction: "In medieval art, thought is the first thing, execution the second; in modern art execution is the first thing, and thought the second. And again, in medieval art, truth is first, beauty second; in modern art, beauty is first, truth second. The medieval principles led *up* to Raphael, and the modern principles lead *down* from him."[27]

By negating the value of performance which operates on the so-called classical ideals of symmetry and regularity aimed at perfect craft, Ruskin builds a new model to be followed, mainly the model of following nature. For instance, the argument of naturalism as a feature of Pre-Raphaelite painting becomes the key one in his writing in defence of the movement: "Pre-Raphaelitism has but one principle, that of absolute, uncompromising truth in all that it does, obtained by working everything, down to the most minute detail, from nature, and from nature only."[28]

In "The Nature of Gothic" Ruskin universalizes his conception of the Middle Ages on the basis of interpreting the artefacts of Gothic art as texts which reveal the meaning he is able to decipher and offer to the reader. Let me give a few more examples of Ruskin's decoding of the culture so as to create the idea of the Middle Ages. In "Modern Landscape" he again opposes the modern to the medieval, this time on the basis of their different way of organizing space:

---

24 Ruskin, "The Nature of Gothic," p. 141.
25 Ruskin, "The Nature of Gothic," p. 143.
26 Ruskin, "The Nature of Gothic," p. 145.
27 Ruskin, "Lecture IV. Pre-Raphaelitism," p. 163.
28 Ruskin, "Lecture IV. Pre-Raphaelitism," p. 169.

> Whereas the medieval was always shutting himself into castles, and behind fosses, and drawing brickwork neatly, and beds of flowers primly; our painters delight in getting to the open fields and moors; abhor all hedges and moats; never paint anything but free-growing trees, and rivers gliding "at their own sweet will"; eschew formality down to the smallest detail; break and displace the brickwork which the medieval would have carefully cemented; leave unpruned the thickets he would have delicately trimmed; and carrying the love of liberty even to license, and the love of wilderness even to ruin, take pleasure at last in every aspect of age and desolation which emancipates the objects of nature from the government of man; - on the castle wall displacing its tapestry with ivy, and spreading, through the garden, the bramble for the rose.[29]

In his recycling of the Middle Ages Ruskin employs an analysis of colour which shows the divergence between the spiritual condition of medieval and modern man. The procedure of significance production starts with showing the impropriety of the name "Dark Ages" as applied to the "medieval centuries:" "They were, on the contrary, the bright ages; ours are the dark ones. I do not mean metaphysically, but literally. They were the ages of gold; ours are the ages of umber."[30]

Despite the claims of literalness in reading the word "bright," when the argument is developed further, colour becomes a metaphor to embrace the spiritual condition of the medieval and the modern:

> For though occasionally glaring or violent, modern color is on the whole eminently sombre, tending continually to gray or brown, and by many of our best painters consistently falsified, with a confessed pride in what they call chaste or subdued tints; so that, whereas a medieval paints his sky bright blue and his foreground bright green, gilds the towers of his castles, and clothes his figures with purple and white, we paint our sky gray, our foreground black, and our foliage brown, and think that enough is sacrificed to the sun in admitting the dangerous brightness of a scarlet cloak or a blue jacket.[31]

Reviewing the idea of the Middle Ages in Ruskin' texts, I cannot resist the impression that the aim of Ruskin's recycling of the Middle Ages is to express his view on modern culture, which functions in contemporary critical reasoning as the Victorian period. The author constructs himself as a representative of his contemporaries or "moderns," as he names them, to present his personal views on some selectively chosen aspects of modern society. What appears in Ruskin's writing is the belief in man as a spiritual phenomenon and in the evolution of

29 John Ruskin, "Modern Landscape," in Ruskin, *Selections and Essays*, pp. 132-133.
30 Ruskin, "Modern Landscape," p. 134.
31 Ruskin, "Modern Landscape," p. 134.

human knowledge as accompanied by simultaneous destruction of his spiritual dimension. The need for religion used to be completed by the culture of Greece or Christian religion of the Middle Ages, yet it cannot be fulfilled by modernity. He writes: "The profoundest reason for this darkness of heart is I believe, our want of faith."[32] Thus the picture of modernity bears hints of a tragic tone. In the words of Chandler, the period of the Middle Ages recycled by Ruskin becomes "a standard by which to measure and modify current-day life."[33]

In "The Pathetic Fallacy," Ruskin claims that the "power of producing sensation" is always "in the thing"[34] and objects to the logics of the division into subjective and objective: "a gentian does not produce the sensation of blueness, if you don't look at it. But it has always the power of doing so; its particles being everlastingly so arranged by its Maker. And, therefore, the gentian and the sky are always verily blue, whatever philosophy may say the contrary; and if you do not see them blue when you look at them, it is not their fault, but yours."[35]

However, the reception of culture is not limited to the reception of sensation. Ruskin's recycling of the Middle Ages as a secondary modelling system is based on certain ontological assumptions, which subordinate the products of culture to some discursive procedures. They are nothing more than selective methods of introducing order by the means of language, procedures typical for a culture which puts stress on the plane of content. For example, the features of Gothic are distinguished within the scope of the previously assumed idea of "gothicness." The scheme is also visible in the placing of the Middle Ages within a chronological order and man within an evolutionary design. On the other hand, the application of discursive methods of ordering is accompanied by the stress on interpreting culture as a collection of signs. In Ruskin the relation between the sign and the signification seems determined. The world becomes a text with Ruskin as its interpreter. According to Lotman and Uspensky this is typical of the culture which emphasizes the plane of expression. *Although his Middle Ages are rich in signifiers, ye, the nature of the signified remains subordinate to the way Ruskin perceives modern culture.* In the analysis of particular signs of culture he is selective, and creates the truth as subjective to the idea that he has in his mind and claims to share with the "moderns:" "I shall only endeavour to analyse the idea which I suppose already to exist in the reader's mind."[36]

Let me close this essay with a quotation from Ruskin which I find relevant to the scope of this discussion: "But above all things, see that you be modest in your thoughts, for of this one thing we may be absolutely sure, that all your

32 Ruskin, "Modern Landscape," p. 135.
33 Chandler, *A Dream of Order,* p. 12.
34 John Ruskin, "The Pathetic Fallacy," in Ruskin, *Selections and Essays*, p. 115.
35 Ruskin, "The Pathetic Fallacy," p. 115.
36 Ruskin, "The Nature of Gothic," p. 139.

thoughts are but degrees of darkness."[37] This comment kept in biblical stylistics is sceptical about the nature of human thought, and in consequence, about language and culture.

37 John Ruskin: "The Relation of Art to Religion," in *Lectures on Art,* ed. George Allen (Orpington & London, 1898), p. 50.

PART III

CHALLENGING REPETITION

Hanna Boguta-Marchel

"Memories are uncertain and the past that was differs little from the past that was not": Some Reflections on the Repetitiveness and Originality of Cormac McCarthy's *Blood Meridian*

Blood Meridian, a novel that can be provisionally designated as a quasi-Western, was published in 1985. Its author, Cormac McCarthy, is presently considered to be one of the most renowned American novelists, though probably still more appreciated by literary critics than by the broader reading public. There is an ongoing dispute, which has roughly divided his critics into two camps, about whether McCarthy's writing is closer to the darkly metaphysical yet altogether affirmative and redemptive tradition of the American South (associated with such figures as William Faulkner and Flannery O'Connor) or to the philosophically broader and more open but at the same time more bitter and "disillusioned" tradition of the West with antecedents in world literature and philosophy (such as Nietzsche, Heidegger, Dostoevsky, or Conrad).[1] Some, therefore, are inclined towards referring to him as a South-Western writer,[2] and *Blood Meridian* is certainly one of McCarthy's most "South-Western" books.

The novel is set on the borderlands between Mexico and the United States, its action beginning directly after the end of the Mexican-American war. It recounts the bloody passage of the – historically factual – Glanton gang of grimly brutish and inhumanly violent outlaws and scalp-hunters who have a contract with local governors to provide Mexicans with the scalps of the daunting Apache who terrorize isolated borderland villages and towns. They therefore simply butcher, in a most forbiddingly cold-blooded and merciless manner, all Apaches they encounter on their way, with time actually developing the habit of killing all, peaceful Indians and Mexicans included, whose scalps they can exchange for pesos. And so the general impression that we get after a first reading of the novel is a quite overwhelming repugnance and irrepressible nausea caused by the excess of surging blood, sizzling brains, pulsating viscera, and the reeking odour of dismembered and rotting bodies. As one critic put it, "One gluts upon a baroque of

1 See Dana Phillips, "History and the Ugly Facts of Cormac McCarthy's Blood Meridian," American Literature 68, no. 2 (1996), pp. 433-460.
2 See for instance Mark Eaton, "Dis(re)membered Bodies: Cormac McCarthy's Border Fiction," Modern Fiction Studies 49, no. 1 (2003), pp. 155-180, where he argues that "the criticism on Cormac McCarthy has almost obsessively tried to identify his predecessors among canonical American writers-round up the usual suspects-and thereby position him in the Southern or Western camps," and that instead he should be viewed "in a different light as part of a new American Studies attentive to the diverse cultural, historical, and literary discourses of the Southwest borderlands" which would reposition his work "within the emergent field of 'postnationalist' American studies."

thieving, raping, shooting, slashing, hanging, scalping, burning, bashing, hacking, stabbing..."[3] Whether the figures who are presented in *Blood Meridian* may be called "protagonists" is a claim of a disputable nature since the narration consistently avoids delving into their psychological disposition, their potential motivation, and their past in general. The narrator seems to treat them as part of the surrounding barren landscape, describing with equally minute details and equally detached impartiality their rotting and reeking wounds, their saddles and weapons, as well as the desolate sands and rocks of the desert they cross. Nevertheless, at least two of these figures deserve closer attention. One of them may be said to function as the "main protagonist" though he, indicatively, does not have a name and is first referred to as the "kid" and then, when the action skips almost thirty years into the future to give an account of the last years of his life, he is labelled the "man." Yet if he is the thread that binds the succeeding events, it is a thread of a very feeble and tattered kind, since he is put forward as the focalizer only occasionally, and his own utterances are practically limited to hostile grunts, shrugging, and spitting into the dirt. Judge Holden, in turn, is a definitely more developed and more compelling character; with his dreadful savagery combined with an extraordinary intellectual capacity, with his enormously tall body with not a single hair on it, his "strangely childlike"[4] face, and surprisingly small hands and feet, he is often alluded to as a figure from another world.

I would like to discuss two motifs of circling and recycling in *Blood Meridian*: the first one concerns a quite literal "recycling" of history in the novel and the second is connected with the issue of artistic creation as a mode of recycling the Real.

One of the diverting facts about the book is that it is based on authentic historical sources. In the only full-time interview Cormac McCarthy ever agreed to give, he said that "The ugly fact is that books are made out of books,"[5] which became a statement his critics hold on to, justifiably or not, as the writer's credo. And *Blood Meridian* is certainly "made out of" a number of such masterpieces as Mark Twain's *Huckleberry Finn* (with a "kid" of similar age leaving an alcoholic father), Faulkner's *Absalom, Absalom!* (again a fourteen-year-old protagonist prompted to search the territory on his own), and Melville's *Moby Dick* (with Ahab's unscrupulous will to power somehow reflected in the figure of Judge Holden, who, for that matter, has about him something of the enormous white whale itself), with Faulknerian strings of adjectives and sentences running on for

---

3 Peter Josyph, "Blood Music: Reading Blood Meridian," in Sacred Violence: A Reader's Companion to Cormac McCarthy, eds. Wade Hall and Rick Wallach (El Paso, Texas: Texas Western Press, 1995), p. 170.

4 Corman McCarthy, Blood Meridian or the Evening Redness in the West (New York: Vintage Books, 1992), p. 6.

5 See Richard B. Woodward, "Cormac McCarthy's Venomous Fiction," The New York Times Magazine, 19 April 1992, p. 31.

two pages without pause, as well as intense and minutely detailed lyrical passages in Twain's style.[6] But *Blood Meridian* also directly refers to historical sources from the second half of the nineteenth century, especially to Samuel Chamberlain's personal narrative *My Confession*. Chamberlain was a soldier who deserted from the US army during the war with Mexico and joined a group of scalp hunters led by John Joel Glanton; the leader of the gang in McCarthy's novel bears the same name. Another significant historical figure McCarthy seems to have drawn from *My Confession* is Judge Holden, whom Chamberlain introduces as "a man of gigantic size" who "stood six feet six in his moccasins, had a large fleshy frame, a dull tallow colored face destitute of hair and all expression" and whose "desires was blood and women."[7] For the sake of comparison let me quote McCarthy's first mention of Judge Holden in *Blood Meridian*: "An enormous man dressed in an oilcloth slicker had entered the tent and removed his hat. He was bald as a stone and he had no trace of beard and he had no brows to his eyes nor lashes to them. He was close on to seven feet in his height..."[8] Chamberlain also notes that "Holden was by far the best educated man in northern Mexico," but also that "a cooler blooded villain never went unhung."[9] McCarthy clearly builds the disposition of his own protagonist on this peculiar association of ruthless violence on the one hand and astounding erudition and eloquence on the other. Chamberlain is also apparently the source for certain specific episodes recounted in *Blood Meridian*, such as Glanton's drunken fit of rage,[10] or instances in which the Judge turns out to be a child molester though no one ever dares to openly charge him with any assault.[11] Collecting scalps or ears of the Indians as "proof" of their capture and a "receipt" on the basis of which the raiders are paid is also a practice that both Chamberlain and McCarthy repeatedly dwell on. Apart from Chamberlain's *Confession*, critics have managed to identify several other historical sources which bear affinity to some of the characters or events described in *Blood Meridian*. These include John Russell Bartlett's *Narrative*, which mentions General Angel Trias, Governor of the State of Chihuahua, and describes him in terms very similar to McCarthy's characterization; John Woodhouse Audubon's *Western Journal* as the source for the Tarot reading tent show which shows up in *Blood Meridian* to join the gang on their way to Janos; and George Fredirick Ruxton's *Adventures in Mexico and the Rocky Mountains* and his account of religious processions flooding

---

6 See Robert Rebein, Hicks, Tribes, and Dirty Realists: American Fiction After Postmodernism (Lexington, Kentucky: The UP of Kentucky, 2001).

7 Quoted in John Emil Sepich, "'What kind of Indians was them?' Some Historical Sources in Cormac McCarthy's Blood Meridian," Southern Quarterly 30, no. 5 (1992), pp. 93-110.

8 McCarthy, Blood Meridian, p. 6.

9 Sepich, "'What kind of Indians was them?'...," p. 97.

10 McCarthy, Blood Meridian, p. 191.

11 McCarthy, Blood Meridian, pp. 118, 160, 191, 239, 275, 333.

the streets of small Mexican towns.[12] The form of McCarthy's novel, with the chapter headings succinctly foretelling the recounted events also seems to allude to authentic chronicles of the period.

And yet, McCarthy's use of all these sources is quite peculiar. First, *Blood Meridian* is decidedly not a historical novel though it does play with the convention by offering a wholly detached, unemotional, and objective account of events practically devoid of authorial judgment or interpretation. Although he apparently put much effort into the study of narratives written by first-hand witnesses travelling in the mid-nineteenth-century Southwest, we clearly sense that McCarthy is not interested in historical accuracy, in dates and factual details. The events he describes and their chronology are only loosely connected with what has actually been chronicled, and he seems to have made no attempt to familiarize the social, cultural, or political context, the whole milieu in which the events take place. In fact, the circle that McCarthy draws apparently takes us back not into the nineteenth century but to much older times, times before any kind of society, culture, or politics were shaped. The protagonists of *Blood Meridian* are not so much historical figures as "beings from an older age"[13]: "Spectre horsemen, pale with dust, anonymous in the renellated heat. Above all else they appeared wholly at venture, primal, provisional, devoid of order. Like beings provoked out of the absolute rock and set nameless and at no remove from their own loomings to wander ravenous and doomed and mute as gorgons shambling the brutal wastes of Gondwanaland in a time before nomenclature was and each was all."[14]

The characters that appear in the novel are repeatedly compared not only to primordial men unaccustomed to any designs of civilization but also to our more primal ancestors, the apes.[15] It therefore seems that finding fault with McCarthy for exaggerating, for overdoing the ferocity of the American raiders, and for replacing the old myth of Western expansion with a new, more gory one (as some of his critics have done) is a misunderstanding. What truly matters in *Blood Meridian* is not the issue of whether the Glanton gang murdered 100 or 1000 Indians; since "memories are uncertain,"[16] as the judge tells the kid, history cannot be fully relied on. What matters is the question what we are actually made of and the uncomfortable recognition that evil and violence are so deeply ingrained in each of us that in particular circumstances they may turn out to constitute our strongest motivation.

There is certainly in McCarthy's novel an element of contemporary historical revisionism of the myth of the Old West, of frontier heroes who fight for

12 For a detailed review of these and other minor sources, their juxtaposition with Blood Meridian, as well as an extensive bibliography, see Sepich, "'What kind of Indians was them?'..."
13 McCarthy, Blood Meridian, p. 176.
14 McCarthy, Blood Meridian, p. 172.
15 McCarthy, Blood Meridian, pp. 90, 153, 200.
16 McCarthy, Blood Meridian, p. 330.

progress and for the expansion of civilization, freedom, and democracy into the remotest corners of the continent. We witness the violence but not the "regeneration" towards which it was believed to lead. Yet *Blood Meridian* lacks the moral overtones typical for revisionist novels; in fact, a number of critics are deeply disturbed by its patent *a*moralism and grieve that such a "highly charged, richly textured novel driven by some of the most impressive American prose of this century features no major figure who is not, quite literally, a slaughterer, and offers scarcely a single act to inspire hope for the race..."[17] McCarthy resists comfortable classification in the revisionist Western history camp since, as Timothy Parrish and Elizabeth Spiller phrase it, his novels recycle the violent history of the Southwest "not to indulge in the compensatory pleasures of self-accusation but to remind us of how particularizing versions of history necessarily deny how we have become to be who we are."[18]

What is also extremely important in the discussion of *Blood Meridian* as a novel that "recycles" a specific moment in the American past is the fact that the history that McCarthy uses is rather unfamiliar to his average reader. He therefore cannot "allude" to it in a typically intertextual modernist manner, but he also does not seem interested in parodying his historical sources by pointing to the gaps and holes in their fabric and by exposing their ethnic bias or their cultural and political entanglement, as a paradigmatic postmodern novelist might be expected to. As Dana Phillips has argued, reading *Blood Meridian* as a parody would be a consolatory way of dealing with the book by making "its text more comfortable, *safer*, than it is"; in fact, "the novel challenges our notions of history and literary history more strongly than 'parody' permits."[19] All in all, *Blood Meridian* does read like a book conscious of its having been "made out of other books," and this awareness renders it more authentic, firm, and reliable, but the novel's relationship to its literary and historical sources is more intricate than it may initially seem. The myth that is revised is much older and more basic than that of the American West; we sense that it touches upon the very "bones of things."[20] The notion that is most strongly challenged early on and consistently through the whole novel is the Rousseauian (and Transcendentalist) belief in man's innate goodness and innocence. At the outset of *Blood Meridian* we face the fourteen-year-old kid, "pale and unwashed," who "can neither read nor write," as he "crouches by the fire" and watches his father who "lies in drink" and "quotes from poets whose names are now lost."[21] His mother died while giving birth to him, and the kid does not even know her name. He soon runs away, first wandering west, then south to New

17 Josyph, "Blood Music...," p. 170.
18 Timothy B. Parrish and Elizabeth A. Spiller, "A Flute Made of Human Bone: Blood Meridian and the Survivors of American History," Prospects 23 (1998), p. 461.
19 Phillips, "History and the Ugly Facts...," p. 458.
20 McCarthy, Blood Meridian, p. 116.
21 McCarthy, Blood Meridian, p. 3.

Orleans. There he manages to survive a truly Darwinian selection deliberately fighting with sailors – men of "all races, all breeds," men "whose speech sounds like the grunting of apes." "They fight with fists, with feet, with bottles or knives." And yet, "the child's face is curiously untouched behind the scars, the eyes oddly innocent."[22] The "taste for mindless violence" that "broods"[23] in him is only triggered by the extreme neglect and lack of love he had experienced, but its roots are deeper and more primordial. These cannot be easily outwitted.

The second "circular" motif of *Blood Meridian* I want to discuss is the theme of artistic creation, which takes us back to the figure of Judge Holden. His creation concerns not only his extraordinary linguistic performances (he repeatedly delivers elaborate speeches on a variety of grandiose topics, such as law, art, metaphysics, war, the cosmos, archaeology, and geology, which present a startling contrast with the apish grunts and mumbles of the rest of the gang), but first and foremost the notes and sketches he constantly makes in his ever-present "leather ledgerbook."[24] As they crossed the desolate mountains and deserts, the judge's habit, even in the most calamitous circumstances, was to stop and collect miscellaneous artefacts with the purpose of copying them into his "little book"[25] and subsequently destroying the originals, so as to "expunge them from the memory of man."[26] He used to stop "watchin the bats,"[27] "botaniz[ing]" and "pressing leaves into his book."[28] When they came upon the ruins of a three-century-old Indian settlement, he "roamed through the ruinous kivas picking up small artifacts" and then "sat upon a high wall and sketched in his book until the light failed."[29] With a telling change of tense (according to my calculations, the narration in *Blood Meridian* changes to the present tense only four times – at the outset and at the ending as well as in the description of two images in the middle, one of them concerning the kid, the other one concerning the judge), the narrator reports:

> In his lap he held the leather ledgerbook and he took up each piece, flint or potsherd or tool of bone, and deftly sketched it into the book. He sketched with a practiced ease and there was no wrinkling of that bald brow or pursing of those oddly childish lips. His fingers traced the impression of old willow wicker on a piece of pottery clay and he put this into his book with nice shadings, an economy of pencil strokes. He is a draftsman as he is other things, well sufficient to the task. He looks

22 McCarthy, Blood Meridian, p. 4.
23 McCarthy, Blood Meridian, p. 3.
24 McCarthy, Blood Meridian, p. 140.
25 McCarthy, Blood Meridian, p. 127.
26 McCarthy, Blood Meridian, p. 140.
27 McCarthy, Blood Meridian, p. 126.
28 McCarthy, Blood Meridian, p. 127.
29 McCarthy, Blood Meridian, p. 139.

> up from time to time at the fire or at his companions in arms or at the night beyond.[30]

When he finished, he "pitched" the artefacts "into the fire."[31] Similarly, when some time later they camped at "the Hueco tanks, a group of natural stone cisterns in the desert," the walls of which were covered with "hundreds" of "ancient paintings," the judge "went among them with assurance" and traced them into his book. "Then he rose and with a piece of broken chert he scappled away one of the designs, leaving no trace of it only a raw place on the stone where it had been."[32]

As Webster, one of his unschooled companions, coarsely points out, "them pictures is like enough the things themselves." Although he highly regards the judge's capacity for faithful representation, Webster forcefully asserts, "But dont draw me ... . For I don't want in your book."[33] He cannot exactly explain the cause of his strong resistance, yet the judge receives it with understanding and tells his comrades gathered around the fire how "he'd once drawn an old Hueco's portrait and unwittingly chained the man to his own likeness. For he could not sleep for fear an enemy might take it and deface it and so like was the portrait that he would not suffer it creased nor anything to touch it ... ." He finally set out on a long journey in search for the judge, found him, and begged him for help. Together they ventured "deep into the mountains and they buried the portrait in the floor of a cave where it lies yet for aught the judge knew."[34]

Judge Holden's strategy of devising exact copies with a subsequent annihilation of their originals is a conscious, carefully calculated scheme which is to grant him a continually expanding orbit of power over nature. As he later explains, in his characteristic forcefully prophetic idiom, "Only nature can enslave man and only when the existence of each last entity is routed out and made to stand naked before him will he be properly suzerain of the earth."[35] And the judge's struggle, in which it is either man who may take "charge of the world"[36] through his power of cognition, or nature, the "smallest crumb" of which "can devour us"[37] if we do not subordinate it first, is a perfectly justified contest considering *Blood Meridian*'s "optical democracy."[38] He is the artist-annihilator, the master of life and death, as he exterminates objects and humans with equal determination, ease, self-assurance, and contentment. He is the proper "suzerain of the earth," creating

30 McCarthy, Blood Meridian, p. 140.
31 McCarthy, Blood Meridian, p. 140.
32 McCarthy, Blood Meridian, p. 173.
33 McCarthy, Blood Meridian, p. 141.
34 McCarthy, Blood Meridian, p. 141.
35 McCarthy, Blood Meridian, p. 198.
36 McCarthy, Blood Meridian, p. 199.
37 McCarthy, Blood Meridian, p. 198.
38 McCarthy, Blood Meridian, p. 247.

exact copies of the Real, signifiers which come to replace their signifieds, signs which are fully mastered by eliminating their referents.

The Judge's "philosophy of representation," if we may call it so, seems to be a radically perverse version of the model of representation which has been termed "Mallarméan" by Michał Paweł Markowski in his brilliant study, *Pragnienie obecności. Filozofie reprezentacji od Platona do Kartezjusza* [*Longing for Presence: Philosophies of Representation from Plato to Descartes*].[39] Markowski, describing the Mallarméan paradigm as one which treats representation as a substitution of the represented object, as its replacement, its "absenting," contrasts it with the "Proustian" model, which, in turn, uses representation as a means of making present that which is at the moment absent, as a chance to participate in the Real that has by now been lost. These two concepts, according to Markowski, lead to two radically diverse notions of literature: literature as an autonomic play of words and literature as a desire to revive reality, and to two contradictory ways of thinking about art in general. Yet both of them presuppose the absence of the represented object (otherwise we would be dealing not with signs but with the things themselves). At the same time, each type of representation would not be possible without an absolutely basic desire, the "longing for presence."[40]

I suppose we may risk the supposition that in the case of Judge Holden this "longing for presence," which for Markowski constitutes the grounds of all art, is lacking since it has been wholly superseded by his longing for power. His art is a "play" in which the "original," autonomous "players" are eliminated and replaced by copies absolutely subordinate to the judge's agency. Apart from being a gifted draftsman and speaker, the judge is also, tellingly, an excellent fiddler, and the others unknowingly dance to his tune: "By now many of Glanton's men were naked and lurching about and the judge soon had them dancing while he fiddled on a crude instrument he'd commandeered … ."[41] In his final divulgence which he presses upon the puzzled kid (now the "man"), he repeatedly uses the metaphor of the dance: "As the dance is the thing with which we are concerned and contains complete within itself its own arrangement and history and finale there is no necessity that the dancers contain these things within themselves as well."[42] The judge is not interested in the dancers themselves as long as their performance observes the rules he himself established.

Interestingly, the narration in *Blood Meridian* frequently highlights the impression that what we are witnessing is only secondary copies of the Real, mere vestiges of the "things in themselves." When, early on in the novel, the kid as a

39 Michał Paweł Markowski, Pragnienie obecności. Filozofie reprezentacji od Platona do Kartezjusza (Gdańsk: Słowo/ obraz terytoria, 1999). See especially pages 7-11.
40 Markowski, Pragnienie obecności…, p. 21.
41 McCarthy, Blood Meridian, p. 241.
42 McCarthy, Blood Meridian, p. 329.

prisoner is taken out into the languid streets of Chihuahua, he observes two copulating dogs: "Two other dogs sat a little apart, squatting loosely in their skins, just frames of dogs in napless hides watching the coupled dogs and then watching the prisoners clanking away up the street. All lightly shimmering in the heat, these lifeforms, like wonders much reduced. Rough likenesses thrown up at hearsay after the things in themselves had faded in men's minds."[43]

When the kid joins Glanton and his companions, the members of the gang are repeatedly illustrated as being devoid of corporeality, as nothing more than outlines or shadows of phenomena that perhaps once had existed for real: "Crossing those barren gravel reefs in the night they seemed remote and without substance. ... A thing surmised from the blackness by the creak of leather and the clink of metal."[44]

What is more, the traces of the Real quickly fade and soon vanish, making it virtually impossible to witness the "things in themselves" at first hand. After their most barbarous massacre, carried out in a village of peaceful, unsuspecting Tigua Indians, where they unswervingly hack "women rising up from their tasks," "old people flinging up their hands," and "children tottering and blinking in the pistolfire," members of the gang leave the ruins of the settlement covered with bodies of the dead who "lay with their peeled skulls like polyps bluely wet or luminescent melons cooling on some mesa of the moon." Yet the narration, shifting into a not so distant future, forecasts that this ghastly horror experienced by so many innocent people will, before long, be remembered by virtually no one: "In the days to come the frail black rebuses of blood in those sands would crack and break and drift away so that in the circuit of few suns all trace of the destruction of these people would be erased. The desert wind would salt their ruins and there would be nothing, nor ghost nor scribe, to tell to any pilgrim in his passing how it was that people had lived in this place and in this place died."[45]

Near the end of the novel, after the gang is scattered in a massacre perpetrated by Yuma Indians, the wounded kid together with the "expriest" Tobin hide away from the judge. The kid uneasily glances at the tracks they left on the sand fearing the judge will follow them. "He looked at the tracks. Faint shapes that backed across the sands and vanished."[46] Since traces disappear so quickly, and since "memories are uncertain," there is no fixed and stable, unquestionable past we may go back to. And yet, paradoxically, *Blood Meridian* takes us back all the time, putting exceptionally strong emphasis on the primitive, primeval nature of its protagonists and events. The result is to depict a state of permanent homelessness, a kind of nomadic uprooting which seem to be nothing modern or postmodern but rather absolutely basic and elemental to our species.

---

43 McCarthy, Blood Meridian, pp. 75-76.
44 McCarthy, Blood Meridian, p. 151.
45 McCarthy, Blood Meridian, p. 174.
46 McCarthy, Blood Meridian, p. 298.

Blood Meridian therefore tells of the recycling of history and the replacement of the original by its feigned copy; history is counterfeited and forgotten, while the Real is brutally annihilated and "expunge[d] from the memory of man."[47] And yet, we sense that the novel may also be read as a search for history, a testimony to an effort to, as Mark Eaton put it, "remember the forgotten … disremembered bodies of those who died in the struggle for survival and territory in the US borderlands."[48] It may too be read as an attempt to revive the longing for representation; even if the judge destroys the Real fearing it may "devour" him, even if his endurance is something "McCarthy's deconstructive approach to meaning and to language seems powerless to resist,"[49] the novel itself is a masterpiece of "Proustian" representation, of making present that which is at the moment absent, and indeed does grant the chance to participate in the lost Real.

47 McCarthy, Blood Meridian, p. 140.
48 Eaton, "Dis(re)membered Bodies…," p. 160.
49 David Holloway, "'A false book is no book at all': The Ideology of Representation in Blood Meridian and the Border Trilogy," in Myth, Legend, Dust: Critical Responses to Cormac McCarthy, ed. Rick Wallach (New York: Manchester UP, 2000), p. 195.

Irena Księżopolska

Recycling the Self: Cultural Amnesia in Michael Ondaatje's *The English Patient*

In Michael Ondaatje's *The English Patient* amnesia functions as one of the main springs of discourse, causing production of a series of narratives as a means to recovery of the principal character's identity. Amnesia is frequently used in fiction and film as a convenient device which allows the possibility of a fresh beginning for the character afflicted with it, a chance to reconstruct his/her self from the position of an impartial observer and readjust this rediscovered entity to the surroundings. However, we could argue that Ondaatje's title character transgresses this convention by not simply recovering and readapting his lost identity, but by recycling the various bits of information left over in his traumatised memory to produce an image of self that is not entirely his own.

Obviously, the English patient is not the typical case of a person suffering from amnesia. On the contrary, his memory seems exceptionally active. "I have always had information like a sea in me,"[1] he tells Hana, and his narratives reveal that he does remember more than is humanly possible, with the gaps seeming to appear only as regards his identity, which, as we soon realize, may be conditioned by his self-preservation instinct, rather than authentic injury.

Is it possible for an amnesiac to retain so much impersonal information in his mind while losing the key to it – the memory of the self? Although it is hard to tell how much Michael Ondaatje was aware of current research in neuroscience while writing his novel, the most recent findings in the field seem to provide evidence for the authenticity of the English patient's condition. Scientists have proposed replacing the long established concepts of short-term and long-term memory with such terms as "feature memory" and "conjunction memory." Research has shown that most amnesiacs do not have problems remembering separate items, but are unable to find the pattern of connections between the images – or, to use a literary term, the plot that links together events and characters. In this light, the patient's claims regarding the "sea of information" as well as his constant attempts to arrange and re-arrange the facts in his stories do not seem inconsistent with the nature of amnesia. His incessant talking would appear to have a therapeutic function, as if he was trying to cure himself by creating a plot to connect whatever isolated items remained in his brain, "piecing individual bits of information together in context."[2] While the patient is able to produce the names of

1 Michael Ondaatje, The English Patient (New York: Vintage International, 1993), p. 18.
2 Ingrid R. Olson, Katie Page, Katherine Sledge Moore, Anjan Chatterjee, Mieke Verfaellie, "Working Memory for Conjunctions Relies on the Medial Temporal Lobe," Journal of

geographical locations and people without hesitation, he seems to have trouble attaching his self to any of the personages he remembers. Hence the frequent switches between first and third person in his narratives – while the patient knows that he did participate in the events he is describing, his damaged memory keeps him on the outskirts of these scenes, as a textual ghost.

Re-plotting his story, recovering it from the ruins left by the great upheavals of public history and personal passion, the patient is simultaneously trying to recover his identity as a teller and a character in the stories.[3] However, the search for his lost self is bound to fail for two major reasons. Firstly, the patient repeatedly states his abhorrence of such totalising concepts as "nations" and ownership and, therefore, notions of belonging. This may imply that even at the beginning of his search his authentic desire is that for self-defeat. He wishes to endlessly explore, but never to mark as discovered the territory of his past. Secondly, the self that he is searching for may be said to exist no longer – whatever he used to be before the plane crash changed during the time preceding his stay in the Villa San Girolamo, and continues to change with the advance of the story. The notion of self as not stable, but as an illusory concept, continually evolving and regressing – "an essentially postmodernist view of identity"[4] seems to be inherent in the novel. Hana and Caravaggio, who used to know each other in the peaceful past, meet almost as strangers, and continue to appear as fascinating strangers to each other throughout the novel. The difference they feel between each other's past and present selves is not imaginary in the least; furthermore, the trauma experience of each makes an uncrossable gap between these two identities. Each of them remembers, but wishes to forget, while the patient seems to be the opposite case, though in fact he only desires to keep remembering, wandering in his deserts, without assembling a memory as a completed narrative. The process of the exploration of Africa, an imperialist pursuit by necessity, becomes mystical because it turns out to be impossible, since the space to be "colonised" is essentially emptiness, and furthermore it turns out to be able to absorb whatever meanings or identities are imposed on it into nothingness – like a black hole of a kind.

There are many signs throughout the text that the coherence of the patient's narrative is illusory and treacherous. The dates seem to be confused at times, and

---

Neuroscience 26, no 4 (2006), pp. 4596 – 4601, Greg Lester, "Lost Connections Amid the Hippocampus: Amnesiac Study Offers Insights into How Working Memory Works," University of Pennsylvania Office of University Communications: 10 Dec. 2006. <http://www.upenn.edu/pennnews/article.php?id=963>.

3 The Villa San Girolamo may be seen as a metaphor of the patient's condition: though it is "a gorge" from the outside, it still contains the magical painted garden, the dangerously mined library with the dormant books waiting to be recovered mirroring the latent memories in the mind of the patient, entwined with dangerous information about his alliances and betrayals.

4 Kristina Stankeviciute, "Constructing the Postmodernist Identity: the Case of Michael Ondaatje's The English Patient," Respectus Philologicus, vol. 3, no 8 (2003), <http://filologija.vukhf.lt/3-8/stank.htm>.

events that were most likely spread out in time are contracted into single entities.[5] For example, let us examine the patient's repeated descriptions of his falling in love with Katharine. He constructs three significantly diverging accounts of this scene. In all three the setting remains the same – the campfire site during an expedition; the patient's attention is drawn to Katharine by her reciting voice. In the version which Hana finds in the diary entry glued into the patient's copy of Herodotus's *Histories*, the event takes place in May 1936, and Katharine is reading a poem by Stephen Crane.[6] However, the patient seems to remember the scene differently when he speaks of it to Hana several pages later – in this version Katharine recites a passage from *Paradise Lost,* one of the key sub-texts of the novel. According to the teller, this takes place within the first few days the Cliftons spent with the expedition, before their return to Cairo for another month of their honeymoon.[7] In the third version, told to Caravaggio, both the date and the passage recited are different: the campfire evening takes place about a month later, after the Cliftons return from Cairo, Katharine appearing to be different (perhaps already harbouring feelings for the patient?), and the passage she chooses to read this time is not a poem, but the story of Candaules and his queen, selected, it seems, to tease her husband, and to entice the patient, read out from his copy of Herodotus's book.[8]

There may be two explanations for this disparity. First, the patient could fail to pin down the exact date of the milestone event because his falling in love was a process, not bound to a single night. The second explanation would be that we are witnessing here the process of a literary revision, a continuous recycling of the basic story into a work of art. Indeed, the last version seems to create the best structural contingencies in the relation to the complete plot, with every detail being strongly motivated, significant, linked to later developments of action. Katharine appears as a seductress, thus releasing the patient from the blame for the affair with his friend's wife.

Another important factor to note as we analyse these diverging yet similar accounts is that in every case the patient falls in love with *the voice* reciting someone else's words. Katharine is giving new life to the written word by lending it her voice – by recycling it from silence, creating it anew for the patient. Simultaneously, listening to the language redeemed through recycling the patient

---

5 For a discussion of this aspect see Amy Novak, "Textual Hauntings: Narrating History, Memory, and Silence in The English Patient," Studies in the Novel 36, no 2 (2004), p. 206, Literature Online: 10 Dec. 2006,
<http://193.0.118.39:8081/han/LiteratureOnline/lion.chadwyck.co.uk/searchFulltext.do?id=R01723721&divLevel=0&area=abell&forward=critref_ft>.
For a general discussion of the problematic sections of the novel see O. W. Pollman, "Canadian Patient: Visit with an Ailing Text", The Antigonish Review, vol. 113 (1999): 10 Dec. 2006 <http://www.antigonishreview.com/bi-113/113-pollmann.html>.
6 Ondaatje, The English Patient, p. 97.
7 Ondaatje, The English Patient, pp. 143-144.
8 Ondaatje, The English Patient, pp. 231-233.

moves out of the circle of the campfire light, thus beginning the process of deleting himself from the discourse.[9]

We may be able to account for some of the textual discrepancies if we remember that the last version is produced under the influence of morphine, fed to the patient in unusually large doses by Caravaggio, who is using the situation to try to discover the identity of the patient. However, the morphine is not a truth potion; it is a drug that may induce euphoric delirium as well as alleviate the pain.[10] In consequence, the confession of sorts regarding his identity which Caravaggio finally manages to *drug-out* of the patient is suspect.

We may, of course, assume that the patient's diary kept before the accident should be considered to be the only truthful account, and that the inconsistencies in the later versions are due to his suspended state of mind. However, further problems are presented by comparing the diary entries with the parts of the text provided by the omniscient narrator. For instance, the consummation of the affair between Katharine and the patient seems to be dated differently. In the diary entry of July 1936 the patient writes about betrayal that comes when "the new lover enters the habits of the other," clearly suggesting that the event has already taken place.[11] However, according to the omniscient narrator, Katharine and the patient become lovers only in 1937, about a year after they meet.[12] This time anomaly cannot be explained by the patient's amnesia, since the diary entry precedes the 1939 accident. The only way to account for it within the logic of the text is to merely accept it as a signal of ambiguity, questioning the connection between Almasy, Katharine Clifton, the patient's diary and the patient himself.

Further doubts regarding the accounts of self provided by the patient arise as we analyse other repetitions in the text. A passage repeated verbatim on two different occasions relates the prehistory of the patient's involvement with Katharine: "In 1936 a young man named Geoffrey Clifton had met a friend at Oxford who mentioned what we were doing. He contacted me, got married the next day, and two weeks later flew with his wife to Cairo."[13] Like all the repetitions in the novel, this one is deeply meaningful, especially since it is a stressed repetition, exact in wording and styled like an introduction to a fairytale. This type of repetition suggests that the fragment is not improvised by the patient, but has been memorized – as part of a poem or an incantation that must precede the ritual of telling the story of Katharine and her lover. With this emphasis, the story that follows must be seen as fictional, created by the teller, not as an account of actual real life events, but as an oral literary production. The patient needs the exact

---

9 Ondaatje, The English Patient, p. 143.
10 "Morphine Sulfate", in: A to Z Drug Facts, (Annapolis, Maryland: Coughlin Indexing Services, Inc., 2005), ClinicalResource@Ovid: 10 Dec. 2006, <http://clinicalresource.ovid.com>.
11 Ondaatje, The English Patient, p. 97.
12 Ondaatje, The English Patient, p. 150.
13 Ondaatje, The English Patient, pp. 142, 229

wording of the overture to remind himself that his story is fictional, even if his listeners miss this clue.

We may also note another oddity in the text concerning the personality of the English patient. While his body is burned beyond all recognition, he somehow retains an amazing charm, drawing to him not only Hana, who clearly sees him as a substitute for her father (who died in the circumstances closely resembling those of the patient), but also the total stranger Kip, and even Caravaggio, who has every reason to treat him as an enemy.[14] He is talkative in the extreme and seems to deeply enjoy companionship. At the same time, all accounts of Almasy in the novel present him as a sullen, quiet and very withdrawn man, who avoids people and behaves without courtesy towards women.[15] There seems to be a marked contrast between the two personalities.

This character metamorphosis can be easily explained if the identity of the desert explorer and German spy Almasy is not that of the patient. He may have been one of Almasy's companions, who came to know the man during the desert expeditions, and became fascinated by this vivid though enigmatic character. In the confusion of the post-crash amnesia, searching for his lost self, the patient may have found it easier to put on the identity of a stranger, known superficially, and therefore appearing as a non-contradictory whole, than to reconstruct his real, conflicting and incoherent identity. The patient's phenomenal memory and knowledge may just make this falsification of identity possible.

A curious confirmation for this theory may be found in the peculiar way the patient emplots the reality around him. If his attempts to talk about his past may be explained as only natural for an amnesiac, it should be noted that his stories have a tendency to stretch further, out of the dimension of personal memories. He seems to feel a constant need of storytelling and is obsessed with decoding the world around him. However, though he claims to possess the knowledge that allows him to reconstruct the map of the world by recognizing its smallest fraction particular to a specific place,[16] in fact, at least on one occasion, he misidentifies his location. He tells Hana that the villa they have taken refuge in must be the famous Villa Bruscoli, thus falsifying its name and plunging on into another maze of stories about Poliziano and his circle, dexterously mixing historical fact with fiction.[17] Moreover, this misnaming may prove to be deliberate. The renaming precedes the storytelling ritual – it is part of the reinvention of the world the patient is busily

---

14 Ondaatje, The English Patient, pp. 91, 89, 252.

15 Ondaatje, The English Patient, pp. 244-245.

16 Ondaatje, The English Patient, p. 19.

17 Ondaatje, The English Patient, pp. 56-57. The story that the patient tells Hana about the beautiful Simonetta, entering the world of preoccupied and mostly homosexual male artists, poets and philosophers and becoming the object of adoration for each of them, seems to mirror Katharine's entry into the exclusively male world of desert explorers, suggesting that there may have been others among them who fell in love with her.

working on. In his seemingly resigned state he still possesses the power to force the real people around him to believe in the phantoms he summons up. The fictionalising of the Villa San Girolamo into the Villa Bruscoli may be a small-scale version of his self-renaming, conducted with unhurried care and definite purpose.[18]

Further evidence for our hypothesis is contained in Caravaggio's story. As the agent of hermeneutic code, the one who seems to solve the enigma of the text, Caravaggio's reliability and authenticity of the motives behind his persistence in decoding the patient's identity are quite crucial. Caravaggio is a thief, a man who lives like a shadow, always slipping through other people's houses. Even though, unlike the patient, he has got a name, it is an absurd one – David Caravaggio, as if the Italian painter and the Biblical king who was the subject of one of his paintings lent the thief their names for a temporary disguise.[19] His personality is obscured rather than defined by his addiction to morphine, obsession with the patient's identity and his war trauma, psychological as well as physical. Ondaatje's readers know this character from the earlier novel *In the Skin of a Lion*, where in the pre-war context he has also appeared as a fugitive, illusory figure. For instance, in one episode he escapes from prison by being painted blue and thus becoming invisible to the guards – literally deleted, painted out of the landscape and therefore surviving.[20] Caravaggio is not only obsessed with identifying the patient, his other constant preoccupation is to remain unrecognised, unidentified, anonymous. His trauma may be said to be caused by this need, as Caravaggio the spy is captured in the flesh because he was first captured on film – he has to steal the photograph in which he was accidentally included and is caught in the act. Caravaggio, the

18 For another instance of a clearly deliberate misidentification see Ondaatje, The English Patient, 154: when Katherine teasingly suggests in front of others that a mark on the patient's forearm is what it is in reality – a woman's bite, he hurries to provide a scientifically feasible cover-up: "It was a scorpion, he said. Androctonus australi." The correct Latin term functions here as a legitimisation of the lie, just as his authentic stories of the Renaissance bohemians uphold the fictitious name of the villa. The use of temporal irregularities in the narrative serves to somewhat obscure this analogy: the scorpion incident, chronologically preceding the villa (mis)naming, is narrated later.

19 Ondaatje, The English Patient, p. 116. It is worthwhile to note that the patient mentions Caravaggio's painting David with the Head of Goliath, describing it as the double self-portrait of the artist. However, what he fails to mention is the fact that there are actually three paintings on the theme attributed to Caravaggio, two of them specifically fitting the description provided by the patient: "the young warrior holds at the end of his outstretched arm the head of Goliath, ravaged and old." The main difference between the paintings is in the face of David – expressing triumph in one (as a warrior saint) and a mixture of disgust and compassion in the other (despairing saint). All the paintings are supposed to have been done late in the artist's life, though, like most of Caravaggio's work, they are undated and unsigned by the artist, and authorship as well as precise dating is still disputed by art critics. This may serve as another signal of the problematic nature of names and dates for signification. As a double self-portrait the painting also hints at the possibility of multiple readings of a single identity. For more information see Howard Hibbard, Caravaggio (Boulder, CO: Westview Press, 1983), pp. 265, 331-332.

20 Michael Ondaatje, In the Skin of a Lion (Ontario, Canada: Penguin Books, 1988), p. 180

named, encumbered by painful memories, is a reversed version of the English patient, his negative, his opposite in most respects, sharing only the central point – the fear, the ultimate unwillingness to be identified, recognition being the thief's ruin. As the patient's double Caravaggio becomes a troubled detective, since by recovering his double's identity he tries to simultaneously delete his own: "But here they were shedding skins. They could imitate nothing but what they were. There was no defence but to look for the truth in others."[21]

Threatened with recognition, Caravaggio uses the patient to bounce the light off himself, to slip back into darkness. He is desperate to remain out of sight and therefore is capable of painting over the blank that the patient represents with the blue colour of invention: "He watches the man in the bed. He needs to know who this Englishman from the desert is, and reveal him for Hana's sake. Or perhaps invent a skin for him, the way tannic acid camouflages a burned man's rawness."[22] What's more, with his spy training, he is well prepared for what secret services of the world call *creating a legend*[23] – a cover-up story that would fictionalise real people and events to mislead the enemy and prevent detection of the agent: "Working in Cairo during the early days of the war, he had been trained to invent double agents or phantoms who would take on flesh. He had been in charge of a mythical agent named 'Cheese,'[24] and he spent weeks clothing him with facts, giving him qualities of character – such as greed and a weakness for drink when he would spill false rumours to the enemy."[25] The quoted passage clearly suggests that the story Caravaggio and the patient collaborate to create is merely a *legend*, a mystification aimed at appeasing and confusing the other side – the reader.[26]

---

21 Ondaatje, The English Patient, p. 117.
22 Ondaatje, The English Patient, p. 117.
23 See Encyklopedia szpiegostwa (Warszawa: Oficyna wydawnicza SPAR, 1995), pp. 146-147.
24 The war of the British and German secret services that constitutes the background of the novel's action deserves further investigation, as it may enhance our understanding of the text by virtually reversing the reading of some of the central characters, in particular that of Katharine. Seen from the perspective of the secret service, Katharine as the wife of an agent participating in an action would have to be an agent herself. If so, her affair with Almasy and its termination may acquire an entirely different meaning. She may have been used by the intelligence to pump the secretive and withdrawn Almasy for information, and when this proved to be ineffective was simply ordered to drop him and find a more productive source. The spy background is all the more interesting due to the fact that Ondaatje evidently drew on factual sources for most of his story. For instance the "Cheese" mentioned in relation to Caravaggio was in reality the code name of the Abwehr's agent network in Cairo, comprised of double agents and used by the "British strategic deception outfit to pass misleading information, on an almost daily basis [...] to the Abwehr's HQ in Rome" - Saul Kally, The Lost Oasis: The Desert War and The Hunt for Zerzura. The True Story Behind "The English Patient" (Boulder, Colorado: Westview Press, 2003), p. 232.
25 Ondaatje, The English Patient, p. 117.
26 In addition, Caravaggio's logic often becomes haphazard since it is fuelled by the morphine he feeds on. For instance, the first time he speaks to Hana of his suspicion that the patient is Almasy, he bases his guess on the name the patient suggests for the dog – Delilah (Ondaatje, The English Patient, pp. 164-165). According to him, Almasy led the German spy Eppler into Cairo and

The patient's strange behaviour towards the English officials who come to question him regarding his past may also indirectly support our hypothesis. If, as his confession to Caravaggio seems to suggest, the amnesia was feigned from the beginning as a kind of self-protection, his way of addressing the suspicious English is surprisingly daring: "You should be trying to trick me... make me speak German, which I can ..."[27] Is he challenging the English to assume he is an enemy and kill him? Burned beyond recognition, tortured by pain and alienation from his past, the patient may be wishing for a quick death. If so, knowing Caravaggio's alliances, he may just as well be teasing Caravaggio to release him from suffering with his counterfeit confession. Some of the words the patient lets slip seem to compound this idea. A comment such as "So you have run me to earth"[28] seems to be of a mocking, teasing nature, and calling the German intelligence officer Rommel "a brilliant man" may be a provocation.[29] If we allow that the patient is secretly courting death as he speaks, his last narratives, and his "confessions" in particular, become even less convincing. We must also not forget the largely negative emphasis put on the word "confess" – it is the word that is used by the Germans when they torture Caravaggio, and that subsequently releases an irrational and powerful fear in him in the totally benign circumstances (e.g. during Hana's game with Kip).[30] Confessing also means tying oneself up with one stable concept of identity, belonging to one nationality, country, class and personality, something the patient particularly wishes to avoid. There may be a deep irony hidden behind his letting Caravaggio attach an incorrect name to his burned legend, if this way the patient avoids being truly recognized.

While we may choose to disbelieve Caravaggio's version of the patient's identity, it is not possible to solve the enigma of the text any other way on the evidence provided in it. There are virtually no leads as to who the patient may be other than Almasy, and in consequence it is not possible to come up with any theory that would reconcile all the contradictions contained in the narratives. One of the few clues may be found, perhaps, in the following bizarre exchange between the patient and Caravaggio: "[C:] 'There were some people in the Intelligence who knew you personally.' [EP:] 'Bagnold probably.' [C:] 'Yes.' [E.P:] 'Very English Englishman.' [C:] 'Yes.'"[31] In a novel entitled *The English Patient* such exchange cannot be accidental. Bagnold, the describer of the dunes, may be the English

disappeared immediately afterwards. Later, the British arrested the spy with the help of an agent with the code name Samson. While Caravaggio himself immediately associates the name Delilah with that of the agent, therefore concluding that the patient's mind works along the same lines, he forgets that even according to his own story Almasy was already gone when Eppler was arrested, and therefore could not have known about the British agent.

27 Ondaatje, The English Patient, p. 95.

28 Ondaatje, The English Patient, p. 252.

29 Ondaatje, The English Patient, p. 254.

30 Ondaatje, The English Patient, p. 223.

31 Ondaatje, The English Patient, p. 255.

shadow that could have been the patient's past, but there is simply not enough information in the novel to develop this hypothesis.[32] Another possibility is Bermann, who along with Almasy has been searchings for the mysterious Zerzura, and who is also glimpsed quoting Herodotus, the book that functions as the index of the patient's forgotten self.[33] Once again, the theory falls flat for lack of evidence. We may be sure, however, that the patient was employed as a spy during the war, though it is not clear on which side. The evidence for this is his extraordinary knowledge of weaponry, too extensive even for a professional soldier, and certainly incongruous with his occupation as a desert explorer.[34] It seems obvious that the secret of the patient is not meant to be solved one way or the other, since the novel "asks both the characters and the reader to reconsider history, the events of the past, from [the] position of the specter."[35] The text focuses on the silences, absences, and the ghosts of history, thus formulating a post-colonial version of events that by necessity avoids the totalising finality of judgments particular for Hegelian dialectics.

The patient's act of falsifying his identity could be explained by his unwillingness to return to a single self, confined within his name – obviously, he longs to stay in the limbo where no name is required, to remain suspended between actual events that he does not remember and the fictional ones he remembers too well. The tactics of storytelling that the patient employs are evidently connected with his desire to remain within the utopian land of no time that he helps create. The villa, misnamed by him, exists as a paradisal island within the sea of destruction, protected as much by its false name (which deletes it from the maps), as by its appearance as a ruin, and by its own legend – a villa with a ghost in the garden.[36] The patient embodies the villa's legend; he becomes the narrating ghost in the artificial garden of his room – a voice within a painting.

His story is already in the past and the fact that he does not fully remember it is the only excuse for his continuing to remain among the living. The patient's story-telling becomes the means by which he continues to exist, somewhat like Scheherazade: he lives only as long as he can continue to spin out his narrative, prolonging it by temporal irregularities as well as by the various omissions. The moment the story completes itself the English patient's life loses its meaning and ends. Significantly, this instant is left unnarrated in the text, though from the very

---

32 Bagnold's name was mentioned earlier by the patient, however, according to his previous statement, Bagnold's involvement with the desert exploration ceased after 1932 – prior to the patient's affair with Katharine (Ondaatje, The English Patient, p. 138). In this light, identifying him as the secret informer during the affair seems inconsistent, adding another shade of ambiguity to the already suspect conversation. Bagnold may be the person behind the code name of Samson, mentioned by Caravaggio (Ondaatje, The English Patient, p. 165).

33 Ondaatje, The English Patient, p. 140.

34 Ondaatje, The English Patient, pp. 20-21, 88-89.

35 Novak, "Textual Hauntings," pp. 206-231.

36 Ondaatje, The English Patient, p. 28.

beginning the reader knows that it is unavoidable.[37] The patient's death is merely a shadow behind his last movements, a shadow he senses in the room, as he decides that both light and sound are no longer needed.[38] As he has nothing else to tell, nothing more to enlighten, the gifts of speech and sight become unnecessary and the patient seems to be able to magically let go of himself and simply cease to be, undramatically, unhistorically.

Turning his copy of the *Histories of Herodotus* into a palimpsest diary, the patient subjects it to a personalising recycling: the first great historian's narrative, coherent, informative and fantastic at once, becomes partially effaced by the enigmatic private history. Yet a paradox occurs: the private in its turn begins to efface itself, deleting the identity of its dramatic personages, becoming once more generalized, impersonal (with the use of the third person the patient adopts after the rupture with Katharine) and finally reduced to a public property which English officials analyse as the patient's only surviving ID, and which strangers read without even asking him for permission.[39] Thus the self-deletion of the patient predates his accident and the loss of memory.

Moreover, the portrayal of self-recycling is not limited to just one character in the novel. Every person in the villa is in some sense displaced, lacking attachment to any culture in particular, or artificially attached to a foreign culture. As has been suggested above, Caravaggio functions as a patient's double, using morphine to become invisible to himself. Hana is another fugitive from history, seeking to reduce her vulnerability by erasing her old personality.[40] Kip is renamed by an alien culture and is described as possessing the ability to become invisible through silence.[41] Therefore, the unifying feature of all four characters in the novel is their loss of identity, the trauma resulting from being burned by whatever elements rage around – war, the deaths of loved ones, betrayal, human savagery, the subtle brutality of nature. Thus cultural amnesia – deliberate or subconscious detachment from one's culture – characterises all the inhabitants of the villa. This detachment from memory constitutes the utopian paradisal niche in which the four

---

37 Ondaatje, The English Patient, p. 4.
38 Ondaatje, The English Patient, p. 298.
39 Ondaatje, The English Patient, p. 97.
40 Hana's spectral nature is also manifested in the fact that she is never given her full name – her surname is not mentioned in the novel. Although Hana also features in Ondaatje's earlier novel In the Skin of a Lion, it is still not possible to deduce her surname. It may be Lewis, after Patrick Lewis, who is identified as her father in The English Patient, but is clearly her step-father from the context of In the Skin of a Lion. She may also be Hana Gull, after Alice Gull, her mother. Another possibility is that Hana kept the name of her actual father, who was killed before her birth, and who is only identified by the fictitious name Cato. One more possibility suggested obliquely in the text is that she may be the daughter of Nicholas Temelcoff. Of the four options listed above two hold additional ambiguity – Cato is only a conspiratorial name disguising the character's real identity, and Alice Gull is a name mockingly adopted by Hana's mother after she narrowly avoids death. The ambiguity of Hana's full name is reinforced in The English Patient by its total omission.
41 Ondaatje, The English Patient, p. 200.

take refuge. They long for and temporarily achieve existence outside of time, race and history, becoming spectral beings, not quite saints, and not angels, but four ghosts in the garden of the partially destroyed Eden.

However, beneath the desire to forget is a repressed but always present need to remember and be remembered. Their amnesia is therefore a re-channelled nostalgia, a yearning for the lost past in which they could have considered themselves as belonging to a culture, and which, of course, is mythical, since when each attempts to analyse the past there is bound to be the same image of failure, of otherness. It can be argued that inability or refusal to remember is a feature characteristic of those whose past is – or seems to be – irrecoverable. Kip and Hana at the end of the novel are able to leave the suspended time zone and slip back into their cultures even though they cannot return to the exact replica of their pre-war selves, being altered not only through the pain of their war experiences but also through their spell of living without memory – through their attachment to each other. They are like ghosts deciding to come back to life, to be reincarnated. Caravaggio's case is more problematic, since, like the patient, he is severely mutilated and has no past to return to. Therefore, Caravaggio also disappears from the book – the last time we see him he is suspended in mid-air, rope-walking to the next villa (probably to rob it) and, as Hana ceases looking at him, simply dissolves into the rain.[42] His affinity with the patient is reinforced this way, and we may read this double disappearance as a suggestion that the two achieved their goal of self-deletion from life as a narrative, managing at the same time to avoid death.

If the story of the patient's selfhood reconstructed in the novel is false, he represents a figure of a doubly recycled personality – as a man who adopts another's story, already containing recycled versions of culture. Paradoxically, at the same time the patient may be seen as an extreme case of originality. By remaining undeciphered and nameless, the patient is virtually a blank, an empty space with no value, just as the desert he pretended to map. This emptiness cannot be said to repeat anything and cannot be repeated. Thus by deleting himself out of existence, by replacing his story with another's, the patient seems to achieve absolute originality. At the same time, as he well knows, there is no such thing as emptiness - the desert is never actually empty – it is full of silenced histories, lost nations, unrecorded passions, secret hiding places and enchanted wells. The diversity underneath the sand dunes is invisible only as long as its spectators are unaware of it, unable to decode it. The English patient's blank originality is merely a mask for the entangled public and private pasts, furtively revising themselves, endlessly recycling the bits of culture surrounding him – passages from books, his own writings, paintings, sculpture, architecture, historical figures and their ghosts,

42 Ondaatje, The English Patient, p. 297. The double disappearance of the patient and Caravaggio from the text without actually "dying on stage" may invite a variant reading in which they both would figure as fantastic constructs invented by Hana to deal with her double trauma (the loss of a father and a child). They move out of the text the moment their mission is accomplished.

real or fictional characters, "interesting geographical problems" [43] of the earth's surface.

Caravaggio, as the patient's reversed double, mirrors his ambivalent position. On the surface of things, for him as a professional thief the appropriation of another's property (a sort of material plagiarism) is a natural habit. His curiosity about other people's lives and customs, while often getting in the way of his professional success, makes him almost a compulsive imitator.[44] At the same time, he is cast in the novel as a man who never reads a book, scorns art, and generally manages to avoid accumulating knowledge, though he seems excellent at finding its sources. Even in his pursuit of the patient's identity he restricts himself to simply talking to the patient, never attempting to analyse the only document that would lend proof to his hypothesis – the patient's volume of Herodotus. Thus, in his own way, he is, like the patient, the pursuer of knowledge in a flux, in the form of transient companionship with strangers, chance encounters and impressions, and not in its final, composed and confined form as a map, book or picture. Once he finds what he thinks is the truth of the patient's story, Caravaggio easily absolves himself of having to remember it by abstaining from judging the patient. Caravaggio as imitator manages to evade recycling by dismissing memory, which allows him to work his vanishing trick, remaining in the flux of things, never solidifying, and therefore staying transparent, inseparable from the background.

For the two characters who "survive the narrative," Hana and Kip, the experience of cultural amnesia has an altogether different meaning. For them, the return to their respective cultures is made possible precisely because they have been forcefully pulled out of the process of automatic recurrence. Their painful experience of cultural detachment becomes the pivotal moment of emptiness that gives meaning to their return to the otherwise mechanical repetition of the daily life, turning it into recycling – a process of repetition with a difference, enriching the texture of life.

---

43 Ondaatje, The English Patient, p. 134.

44 Ondaatje, The English Patient, pp. 169, 208. This tendency to imitate other people's customs is vividly demonstrated in the episode with the robbery of an Indian house, in which Caravaggio, intrigued by the sarongs his awakened hosts wear, tries one on and is almost trapped by it.

Jacek Mydla

Recycling the Spectre: James Boaden's Stage Adaptations of the Gothic Romance and the Spectres of Literary Appropriation

Authorship is a mania, to conquer which no reasons are sufficiently strong.

Matthew Gregory Lewis, *The Monk*

The effulgence of Gothic, its rise and then its rapid degeneration allow us to examine some of the mechanisms that initiate and govern the transmission of literary material in a larger cultural context. Circulation of elements constitutive of the Gothic among the three major Gothicists in the last decade of the eighteenth century, Ann Radcliffe (1764 – 1823), Matthew Gregory Lewis (1775 – 1818), and James Boaden (1762 – 1839) offers examples, not only of literary appropriation, whereby various motifs, themes, and devices are transferred from an early work to its progeny, but also of a somewhat paradoxical joint operation of two conflicting interests, that of outdoing (out-Gothicising) the predecessor and that of correcting his or her work in terms of its adherence to a preconceived ideal of a well-made piece of fiction or drama. To bestow a leitmotif on the analysis to follow, we shall concentrate on the handling of the artistically tricky and morally suspect element of the supernatural. This narrowed perspective will let us study the forces involved in transmissions, or the "recycling," of the literary material in that interesting period when the declining Enlightenment was making a fussy exit with romanticism impatiently awaiting its cue in the wings.

In order to set in perspective Boaden's adaptations of the famous Gothic romances for the London stages, we shall first look briefly at the dissemination of the Gothic, especially in its relation to the changing role of the adaptation. This process will be examined in the context of the popularisation, if not the actual emergence, of the notion of literature as entertainment, i.e. of reception, and of the reading of romances in particular, as a common time-beguiling activity or a leisure-managing diversion. We shall make an attempt to see, on the material that Boaden's dramatisations provides, the conflicting aesthetic and moral tendencies with which this particular project is ridden as exemplary of the more general observation that any recycling of literary material is necessarily a combination of retention and deviation.

By the last decade of the eighteenth century, the newly-born literary genre, the novel, had not only been established as a well-respected literary mode but had also been distinguished from the less "respectable," and more "entertaining" varieties of fiction.[1] In his article on the role of the circulating libraries in the rise of the Gothic romance, Edward Jacobs describes the two-way influence between literary production and reception, emphasising that circulating libraries contributed to the categorisation of fiction as well as functioned as publishing houses. The latter role allowed them much more strongly to influence the process of setting up the Gothic romance as the hottest merchandise on offer in the literary market of the last decades of the eighteenth century.[2] It is by no means a coincidence, argues Jacobs, that the two would-be celebrities of female fiction in this period, Frances Burney and Ann Radcliffe, made their debut thanks to promoting activity of circulating libraries, Thomas Hookham in the case of the latter.[3] Having anonymously had three of her novels published by Thomas Hookham, Radcliffe signed a lucrative deal with G. G. and J. Robinson for the forth one. *The Mysteries of Udolpho*. Subtitled *A Romance* and published in 1794, *Udolpho* liberally employs the devices (such as that of the explained supernatural) which the authoress had successfully used in her earlier fictions, all of which soon were to earn her the somewhat ambiguous title of the "great exemplar" or a model to a train of followers.

The proliferation of the Gothic romance exemplifies and epitomises the processes which were taking place in the vibrant cultural environment of England at the turn of the eighteenth and nineteenth centuries. According to some estimates, publications of Gothic fiction rose from, roughly, one per year in the 1770s and 1780s to over twenty in the 1790s.[4] Naturally, this self-propagating capacity aroused anxiety. From its inception in the 1760s, the Gothic has raised doubts concerning its respectability; not infrequently biological metaphors were employed to capture the genre's unparalleled fecundity.[5] Perhaps devoid of heuristic value, such metaphors indirectly indicate the Gothic's parasitical nature, a cancerous growth, as some would have it, which grafts itself upon the existing literary modes and conventions and makes them propagate with wanton vigour. In his "Advertisement" to *The Monk*, Matthew Gregory Lewis openly admits that he has drawn upon a great number of different sources and suspects that he may be guilty of "plagiarisms" that he is not aware of.

As we have already suggested, the Radcliffe-Lewis-Boaden circle of influence provides model examples of literary recycling: of appropriation, reuse

---

1 See J. M. S. Tompkins, The Popular Novel in England 1770-1800 (London: Methuen & Co Ltd, 1969; 1st published 1932), passim.

2 Edward Jacobs, "The Anonymous Signatures: Circulating Libraries, Conventionality, and the Production of Gothic Romances," English Literary History, 62/3 (1995), pp. 603-629.

3 Jacobs, "The Anonymous Signatures," pp. 620-621.

4 See Robert Miles, "The 1790s: the Effulgence of Gothic," in Jerrold E. Hogle, ed., The Cambridge Companion to English Fiction (Cambridge: Cambridge University Press, 2002), p. 43.

5 See Elizabeth R. Napier, The Failure of Gothic. Problems of Disjunction in an Eighteenth-century Literary Form (Oxford: Clarendon Press, 1987), p. viii.

and reworking, and – perhaps more interestingly – of emulation and competition. Radcliffe's impact upon the developments of the Gothic in fact extends far beyond the influence which her romances exerted upon Lewis and Boaden. As the authoress's biographer, Rictor Norton, has stated using Mary Shelley's metaphor which combines procreation and illegitimacy, "Ann Radcliffe's hideous progeny is enormous."[6] Clara McIntyre, an earlier biographer, remarked that "a lady of any literary conscience might well have a sense of guilt at being responsible for such a following."[7] Norton estimates that "about a third of all the novels published between 1796 and 1806, and many serials in ladies' magazines, had scenes inspired by *A Sicilian Romance* and *The Romance of the Forest*.[8] The number of examples that Norton himself cites is indeed impressive, and gives us an idea of the kind of corruption involved: "Thinly disguised redactions of her books were serialized as shilling shockers, with everything stripped away except the sensational, which thereby appeared even more gross in the absence of Ann Radcliffe's restraining sentiment."[9]

Even though Norton does not mention Lewis, the "lack of restraining sentiment" applies very well to this infamous follower of Radcliffe. Her *Udolpho* may have even saved what was to become the most sensational publication of the 1790s from being consumed by flames. In a letter to his mother written in May of 1794, Lewis confesses to his fresh inspiration: "I have taken up my Romance, and perhaps by this time Ten years I may make shift to finish it fit for throwing into the fire. I was induced to go on with it by reading 'the Mysteries of Udolpho, which is in my opinion one of the most interesting Books that ever have [sic!] been published."[10] Among Radcliffe's followers, Lewis is distinguished by his youthful – and irresponsible in a soon-to-become MP – ambition to out-Gothicise the great exemplar, or, to use the terms applied later in Radcliffe's essay on the supernatural in poetry, to supply horror in the place of terror: by making the genre more spectacular and more sensational ("unpardonable grossness" in the words of Hazlitt[11]), Lewis certainly contributed to its "degeneration," and, for the same reasons, to its more vigorous proliferation.[12] "The record of its [*The Monk*'s] reappearances is altogether remarkable for a Gothic romance," writes Louis Peck

6 Rictor Norton, Mistress of Udolpho. The Life of Ann Radcliffe (London and New York: Leicester University Press, 1999), p. 163.

7 Clara F. McIntyre, Ann Radcliffe in Relation to her Time (1920), quoted by Norton, Mistress of Udolpho, p. 163. A contemporary reviewer wrote about The Mysteries of Udolpho that the book "had given birth to several humble imitations." (ibidem, p. 162).

8 Norton, Mistress of Udolpho, p. 162.

9 Norton, Mistress of Udolpho, pp. 162-3.

10 From Lewis' "Selected Letters" published as part of Louis F. Peck, A Life of Matthew G. Lewis (Cambridge, Mass.: Harvard University Press, 1961), p. 208.

11 From Hazlitt's lectures of 1819; quoted in Peck, A Life of Matthew G. Lewis, p. 37.

12 Degeneration is a term used by many critics, also, predictably, by Napier; see The Failure of Gothic, p. 31.

about Lewis's novel, and goes on to quote the impressive number of editions (in England and the United States), translations (French, German, first, and then Italian, Spanish, Swedish, and others), and so-called chapbook versions for the circulating libraries.[13]

Of special interest is, however, the fact that *The Monk* displayed, in the words of Peck, "Lewis' instinct for dramatic presentation." The critic cites a number of examples of this skill, i.e. of the way in which Lewis dexterously exploited the mimetic potential of his story. This is how Peck compares Lewis's with Radcliffe's type of Gothic (apparently with the intention to disparage the authoress): "caring nothing for [Radcliffe's] long descriptions filled with romantic vagueness intended to evoke an atmosphere of mystery, Lewis arranges his scenes as for a stage spectacle with gaudy colors and striking contrasts of light and darkness."[14] Peck chooses not to mention that what added to the scandal of Lewis' Gothic was that the supernatural was allowed full manifestation. We might say that Lewis' horror Gothic came to fulfil those expectations which were raised in romance readers by Radcliffe's technique of explaining the supernatural away. The difference between the two authors can also be described thus: Radcliffe's type of the Gothic is diegetic while Lewis's is mimetic. Lewis's version of the Gothic offers dramatic vigour, and graphic literalness instead of Radcliffe's suggestiveness, obscurity, drawn-out suspense, and postponed revelation. In approaching both these immensely popular authors in the hope of extracting theatrically viable material, Boaden found himself hard pressed to negotiate two very distinct varieties of the Gothic, the so-called terror and horror modes.[15]

While influence, as we go from *Udolpho* to *The Monk* and from *The Monk* to *The Italian*, seems to be a matter either of conjecture or of fine comparative study, Boaden's adaptations openly reveal their credits. *Fointainville Forest* (1794) is "founded on the Romance of the Forest"[16] while the very title *The Italian Monk* (1797) cunningly combines the titles of the two source romances while being actually "founded" upon Radcliffe's. The title *Aurelio and Miranda* (1798) conceals the source narrative, but the list of characters makes it clear that Boaden simply decided to change the names of Lewis's Ambrosio and Matilda, whom he has made the protagonists of his melodramatic version of *The Monk*, on which, as the Advertisement informs us, "the Play is avowedly founded." However, no

---

13 Peck, A Life of Matthew G. Lewis, p. 37.

14 Peck, A Life of Matthew G. Lewis, p. 42.

15 On this distinction, which derives from Radcliffe's essay "On the Supernatural in Poetry" (published in 1826), see David S. Miall, "Gothic Fiction," in Duncan Wu, ed., A Companion to Romanticism (Malden, Mass.: Blackwell, 1999), p. 349.

16 All citations to Boaden's plays are to The Plays of James Boaden, edited with an introduction by Steven Cohan (New York and London, Garland Publishing, Inc., 1980), which belongs to a series of reprints, Eighteenth Century English Drama. References are by page number (individual pagination to every play) and by act and scene, e.g. FF III/2, 34. The original editions do not affix numbers to scenes.

matter how slim or how glaring a particular piece of evidence of appropriation might seem, we still need some theoretical fine tuning in order, first, to give it an adequate description, and, next, to be able to assess its significance. In our case, we need a working definition of the Gothic, and more particularly a clear notion of the supernatural as a recyclable element of Gothicness.

Critical debates over the Gothic have been ridden with a strange kind of ambiguity; while some have been troubled by the Gothic's generic amorphousness others insisted on its fundamentally formulaic nature. One the one hand, the Gothic has eluded too narrow or precise definitions and has come down as a literary fad which blatantly goes beyond the conventional generic classifications.[17] There is for instance no justification for treating the Gothic exclusively as a mode of fiction. Also, attempts to single out a collection of Gothic features are either nebulous (Evans's "mystery, gloom, and terror"[18]) or unjustifiably reductive. Gothicness seems to spread over a great number of devices, recurrent and reproducible; none of them may be constitutive of the genre, but collectively they allow us to come up with typologies of historically prevalent types of the Gothic, such as the "terror" and "horror" varieties, each made up by a peculiar configuration of devices.[19] On the other hand and leaving aside these considerations, critics seem to agree that the Gothic is a highly conventionalised literary form. Elizabeth Napier, in her study of the "failure of Gothic," has given a framework to concerns voiced by many critics. She regarded Gothicness as a system or "a collection of Ghost-story devices" (after Robert Hume) and described Gothic narratives as highly formulaic.[20] Here are some sample statements:

---

17 Elizabeth Napier, for instance, names the following genres that Gothic combines: fairy tale, romance, Jacobean drama, and novel of manners; such mixing being a case of what the critic calls, after Robert Platzner, "generic instability"; see Napier, The Failure of Gothic, p. 67. Cox speaks of "generic profusion" which is "often accompanied by innovative generic combination and modification"; Philips Cox, Reading Adaptations. Novels and Verse Narratives on the Stage, 1790-1840 (Manchester and New York: Manchester University Press, 2000), p. 6.

18 Paul Ranger denies the possibility of defining Gothic and claims that "it [the gothic] can be thought of as an artistic climate..."; Paul Ranger, "Terror and Pity reign in every Breast." Gothic Drama in the London Patent Theatres, 1750-1820 (London: The Society for Theater Research, 1991), p. 17.

19 For a justifiably concise "definition" of this type see Kelly Hurley, "British Gothic Fiction, 1885-1930," in Jerrold E. Hogle, ed., The Cambridge Companion to Gothic Fiction (Cambridge: Cambridge University Press, 2002), p. 191. Hurley names stock characters and stock events, setting, theme, style, narrative strategies, and "affective relations to its readership" as the basic elements of such an extended definition of Gothic.

20 Napier, The Failure of Gothic, p. 29. The term "formulaic" is repeated three times on this page. Napier borrows the phrase quoted from Robert Hume's celebrated essay "Gothic Versus Romantic: a Revaluation of the Gothic Novel" of 1969.

> [Gothicism] is a standardized, absolutely formulaic system of creating a certain kind of atmosphere in which a reader's sensibility toward fear and horror is exercised in predictable ways.

Napier argues that "the superficial and the formulaic ... form the very heart of the Gothic" and that the Gothic text arises "from the repetition of a certain series of extremely conventional scenes, events and landscapes." By this rationale, a Gothicist makes a conscious decision to employ already existing building blocks, giving them a predictable treatment, i.e. to arrange them according to well-known narrative strategies or "recipes." In other words, if one decides to write another story "of mystery and terror," one joins the ever-growing crowd of literary recyclers.

The conventionality of the Gothic allowed for satirising it by composing recipes. In an anonymous letter on "Terrorist Novel Writing" we read of "the insipid repetition of the same bugbears."[21] But why, one might ask, should Gothic of all literary modes bear the brunt since a lesser or higher degree of the formulaic lies at the bottom of every literary genre or form? The answer, I believe, lies in part in the way in which early Gothic authors styled themselves as Originals. The early Gothic of the Radcliffean variety was predicated upon the idea of the original genius (or the poetic Genius and an Original); Radcliffe aspired to poetry rather shilling shockers and the fact that her fictions generated a spawn of imitations must have been a shock indeed. Radcliffe may not have been personally guilty of having attracted legions of imitators and adaptors (so-called hack dramatists among them); the "fault" seems to lies in the ready-to-use patterns and formulas of the "parent narratives." This in turn demonstrates that authors are "guilty" of having written with the help of identifiable devices and various "tricks of the trade" (such as readers and likely imitators are able to fish out), and especially of writing in order to achieve a clearly stated goal, that of thrilling the reader. This brings us to the problem of the supernatural and the way in which the spectre raised critical debate. On the one hand, a ghost, even if contained in the author's green room, is a perfect realisation of the promise of "terror and pity" on which the Gothic is predicated; this seems to have been perfectly understood by Radcliffe. On the other hand, the fulfilment of this promise logically follows from this premise, which seems to have been perfectly understood by Lewis. And what can be more effective and accomplished than putting the spectre on the stage? Nothing, as the example of Shakespeare's supernatural scenes showed, and which both Lewis and Boaden seem to have comprehended equally well. But what they understood just as well was that spectres were not to be tolerated in an age in which, as Lewis himself put

21 See E. J. Clery and Robert Miles, eds., Gothic Documents. A Sourcebook 1700-1820 (Manchester & New York: Manchester University Press, 2000), p. 184.

it, “the belief in Ghosts no longer exists!”[22] Caught in this conflict between artistic effectiveness and moral propriety, Lewis makes an attempt to vindicate the former: “there is now no fear of increasing the influence of superstition, or strengthening the prejudices of the weak-minded” (*ibidem*). Boaden, on the other hand, as we shall see, becomes increasingly cautious and in his last attempt at adapting a Gothic romance, i.e. in *Aurelio and Miranda*, resolved to do “without supernatural agency.” This dispute over the alleged impropriety of spectres gliding across page or stage, however, ought not to blind us to the fact that the supernatural, explained away or not, is an indispensable element of the Gothic formula. Thus, if an author decides to use the recipe, he or she needs to add this particular ingredient. If Boaden’s plays exemplify this generic necessity, they also help us reveal the authorial anxieties involved.

The Gothic’s proliferation was substantially aided by the genre’s transference to the medium of the theatre. However, Gothic adaptations – in the period under scrutiny – were merely a section, and must be regarded as example, of the growing popularity of the adaptation as a particular type of cultural recycling. In the second half of the eighteenth century a new type of adaptation emerged, a type which has become common nowadays: a novel (or, rarely, an epic poem) is transferred to another medium, one which is more entertaining, more dynamic, more accessible, and more vivid.[23] The increase in the number of adaptations has been put down to a dynamic development of the theatre: more stages and types of entertainment, enlargements of audience space in the Theatres Royal, and the resulting growing number of theatregoers. All this caused an increasing demand for theatrical entertainment, and the outcome was easy to predict: “to fill that demand in rushed the hack dramatists with versions of virtually every recent novel.”[24]

Like the two-way influence between the circulating library and the production of Gothic romances described above, the relation between fiction and theatre was not one between two independent and frozen entities. When publishing his adaptation of Radcliffe’s *The Romance of the Forest*, Boaden inserts a note under “the dramatis personae” in which he first explains that “It was not from a vain tenaciousness that I determined to retain passages expunged in the performance,” and then adds that “The Stage and the Closet are very different mediums for our observance of effect.”[25] This helps us conceive, as a possibility, a ballet pantomime adaptation of a bulky Gothic novel, such as that of *The Monk*. This possibility became an actuality which, under the title *Don Raymond; or the*

---

22 Matthew Gregory Lewis, “Postscript to The Castle Spectre,” in Clery and Miles, eds., Gothic Documents, p. 198.

23 “Spectacular settings, elaborate costumes and the colourful effect of massed gatherings added zest to the gothic drama.” Ranger, “Terror and Pity reign in every Breast,” p. 75.

24 H. Philip Bolton, Dickens Dramatized (1987), quoted by Cox, Reading Adaptations, p. 1.

25 Boaden, Fontainville Forest, page not numbered.

Castle of Lindenburgh, was presented at Covent Garden in March 1797.[26] Boaden's decision to get the text of his play published in a non-expunged version might testify to his desire to attain similar "effects" as those of the source narrative, i.e. "the effects of the Closet," as he calls them.[27] Taking this into consideration, there was a double purpose to his adaptations: they were a literary product capable of realising the effects of both drama and fiction. To complete the picture, Boaden's *Fontainville Forest* is published in 1794 by the same circulating-library publisher, Thomas Hookham and Carpenter, as was its fictional prototype, *The Romance of the Forest* three years earlier. Furthermore, Boaden's marginal remark makes us realise the divide which separated the private and the public manners of consuming literary material and the conventions which governed this consumption. The ban on the supernatural can be perceived as motivated by a conviction that the stage gives public sanction to the supernatural, which in the case of Lewis' spectres of German descent was politically suspect.

Theatrical adaptations of early Gothic fiction have to be regarded in the thus delineated context, which, among other things, involves the idea of literary degeneration. Reflecting this, scholars have repeatedly used deprecatory metaphors, e.g. that of "looting." According to Allardyce Nicoll, "the whole field of fiction was eagerly and systematically sacked" by "hack" playwrights.[28] At the dawn of high romanticism, when poetic aspirations soared, many lamented the, in the words of Wordsworth, "depraved state of the stage."[29] Wordsworth believes that in rendering an adaptation, the Dramatic Author "adapt[s] himself to the taste of the Audience"; by this rationale, true poetic greatness, such as that of Shakespeare, consists in not "stooping to accommodate himself to the people."[30] If the theatre was a suspect medium due to the perceived lowering of standards of taste then what could have been worse that the staging of a Gothic romance? Yet this is exactly where the Gothic was headed in the radicalised, "horror" version. Indeed, Lewis's dramatic instinct was soon to find an outlet in a successful theatrical debut, *The Castle Spectre* (Drury Lane, December 1797).[31] A landmark Gothic drama if judged by its popularity, the merits of the play do not transpire "upon reading."[32] This is hardly surprising: the play successfully transfers

26 See Bertrand Evans, Gothic Drama from Walpole to Shelley (Berkeley and Los Angeles: University of California Press, 1947), p. 241.

27 These effects are of course narrated in the MS-reading episode in The Romance of the Forest itself, which adds to this novel an interesting meta-fictional dimension.

28 Allardyce Nicoll, A History of English Drama 1660-1900 (1955), quoted by Cox, Reading Adaptations, p. 1.

29 Quoted by Cox, Reading Adaptations, p. 19.

30 Wordsworth, Essay Supplementary (1811), quoted by Cox, Reading Adaptations, p. 11.

31 Reprinted in Jeffrey Cox, ed., Seven Gothic Dramas, 1789-1825 (Columbus: Ohio State University Press, 1992). For a reconstruction of the staging see Ranger, "Terror and Pity reign in every Breast," pp. 116 ff.

32 Peck, A Life of Matthew G. Lewis, p. 72.

conventions that govern, if latently, literary Gothicness into another medium, that of the theatre. Lewis's determination to stage a ghost appearance – "I confess I cannot see any reason why Apparitions may not be as well permitted to stalk in a tragedy, as Fairies be suffered to fly in a pantomime"[33] – despite admonitions and warnings,[34] testifies to a self-conscious operation. Symptomatically, as in the case of Boaden's plays, the published text is "considerably shorter" than the published version due to the editing out of soliloquies and dialogue.[35]

Boaden's first adaptation was of Radcliffe's *The Romance of the Forest*, a novel in three volumes, which, it will be recalled, was originally "printed for T. Hookham and Carpenter" in 1791, as we are informed on the title page. Boaden's five-act dramatisation, *Fontainville Forest*, was performed at Covent Garden in March 1794[36] and printed, likewise, "for Hookham and Carpenter" in the same year. The title page pays due credit to the source, informing the reader, as we have seen, that the play has been "founded on the Romance of the Forest," or, according to the Prologue, "caught from the Gothic treasure of Romance."

Fontainville Forest is in a way the boldest of Boaden's three adaptations due to his handling of the supernatural. Boaden decided to have a ghost on the stage, thereby turning into a stage occurrence Radcliffe's suggestions of supernatural presence in Chapters IX and X (vol. II) of the novel.[37] Here Radcliffe narrates the heroine's perusal of the mysterious manuscript, discovered in one of the chambers of the abbey. As our brief analysis will show, in transferring the diegetic material of the source narrative onto the mimetic medium of the theatre, Boaden faced the extra challenge of handling a double narrative, for the story proper has embedded within it the manuscript, which Adeline peruses and to the content of which she responds.[38] Another complication arises from the fact that Radcliffe's narrative relates as many as three "prophetic" dreams which as it were prepare Adeline for the supernatural horrors awakened by the MS. As Boaden decided not to excise this element of the story from his play, he again faced the necessity of negotiating a dramaturgically tricky material, also of a two-tier diegetic structure.

---

33 Lewis, "Postscript to The Castle Spectre," p. 198. Lewis' continued experimentation with the supernatural involved also a representation of the Ghost of the Bleeding Nun; see Ranger, "Terror and Pity reign in every Breast,", p. 76.

34 "Georgian audiences were doubtful about the propriety of the appearance of specters." Ranger, "Terror and Pity reign in every Breast," p. 75. Shakespeare's ghosts were of course the great exception.

35 Peck, A Life of Matthew G. Lewis, p. 75.

36 See Evans, Gothic Drama, p. 240. The title page of the printed version does not name the date of first performance.

37 Ann Radcliffe, The Romance of the Forest (Oxford: Oxford University Press, 1999). All citations, marked as RF plus page number are given in the main text.

38 In other words, the novel narrates the reading of a story "of mystery of terror" while Boaden in his adaptation stages that reading. Boaden's adaptation of the MS-reading episode largely deprives the reading process as depicted by Radcliffe of its radical privacy, essential, as we have already remarked, to the consumption of Gothic.

In the novel, Adeline's prophetic dreams (Chapter VII, which rounds off volume I) precede her perusal of the manuscript (Chapters IX and X). *Fontainville Forest* basically reverses this; in Act II Scene 5, Adeline finds the rusty dagger and the "scroll," which concludes Act II; in the scene which follows (III/1), she reveals her dreams to Madame Lamotte (Radcliffe's La Motte). Leaving aside these differences, the MS-perusal scene (III/4), the final scene of Act III, occupies a central position in Boaden's play. Verbal echoing by means of paraphrase, evidently Boaden's favourite method throughout, abounds throughout and the scenes under scrutiny are no different in this respect. Before stumbling upon the dagger, Boaden gives Adeline the following lines:

> … but asleep, or waking, still
> Conviction haunts me, that some mystery
> Is wrapped within these chambers, which my fate
> Will have me penetrate. – FF, *II/4 (24)*

This corresponds closely to Adeline's solitary musings in the novel in the episode when she decides to peep into the uninviting interior of the chamber: "'A mystery seems to hang over these chambers,' said she, 'which it is, perhaps, my lot to develope [sic]; I will, at least, see to what this door leads'" (*RF* 115). Boaden is clearly drawn to the mimetic potential of the source narrative. He reuses the moment when Adeline speaks out loud; steering clear of direct plagiarism, he paraphrases the words which Radcliffe put in the mouth of *her* heroine.

In the main scene, that of MS perusal, Boaden first of all performs the expected compression of the manuscript, compression being the principal method when a playwright adapts a copious narrative such as Radcliffe's. He still makes Adeline read out the text, but in his version the manuscript is not only shorter but also reveals much more about identity of its author. He introduces himself to Adeline and to the audience already in the opening sentence: "The wretched Philip, Marquis of Montault, / Bequeaths his sorrows to avenging time" (*FF* III/4; 38).[39] Soon enough we find about the fratricide itself: "Yet, O my brother, I had never wrong'd you" (*FF* III/4; 39). In the novel, Radcliffe, faithful to her method of postponed revelation, discloses these basic facts at late as Chapter XXIII, i.e. almost at the close of the narrative.

Three times in the relatively brief scene Boaden has a ghost interrupt Adline's reading. Boaden makes the appearance of the ghost gradually manifest, commencing with (according to stage directions in the printed version of the play which we are forced to rely on) a voice "heard within the chamber," through becoming "faintly visible," to a full-blown parade across the stage: "The phantom

39 Boaden, perhaps unintentionally, confused the names of the two brothers in Radcliffe's story, where Phillip de Montalt is the villain and his brother's, i.e. Henry's, assassin.

glides across the dark part of the Chamber, Adeline shrieks, and falls back. The Scene closes upon her" (end of Act III, pp. 39-40).[40]

As we go through Radcliffe's narrative, we are realise the difficulty which Boaden faced when trying to come up with a successful dramatisation of this scene, one of the most accomplished realisations of Radcliffe's idea of terror. First of all the process of reading, in Radcliffe's rendition of it, is a highly personalised business; the emphasis is put on the effect which the content of the MS has on Adeline's imagination. As a result the readers cannot be sure whether, when Adeline hears the word "*Hear*" "distinctly repeated by a whisper immediately behind her" (*RF* 132-133) this is not an auditory figment of her heated imagination. When some time later she first hears a "hollow sigh" and then thinks she sees "a figure, whose exact form she [cannot] distinguish," "pass along an obscure part of the chamber" (*RF* 134), this again is a moment when her imagination has "refused any longer the controul [sic] of reason." This kind of ambiguity, the achievement of which was the cornerstone of Radcliffe's aesthetics of terror, was basically unattainable on the stage. The decision which Beaden was forced to make pushed him straight into the embrace of the supernatural.

At the end of *Fontainville Forest*, Boaden appends an epilogue written for Mrs. Pope, the actress playing Adeline. As well as providing some self-satirising quibbling, this piece also documents the ambiguous status of the supernatural in the dramatic exploits of the period (see Lewis' Postscript to *The Castle Spectre* quoted above). The dubious status of the supernatural on the stage made some justification necessary. The epilogue makes it clear that the age was not prepared to tolerate any other ghosts than those "of Hamlet's pedigree" (*FF* 69). In reply to the objections raised by the actress, the author seeks to "sanction" his spectre by claiming for it the Shakespearean pedigree and eventually comes up with the following pun: "Madam I die, if I give up the ghost."

Boaden's second and ultimate engagement with Radcliffe's romances was *The Italian Monk*, a play in three acts, first performed at Theatre Royal, Haymarket, on August 15th, 1797, and subsequently published in the same year by the Robinsons. Boaden handles Radcliffe's "response" to *The Monk* (for apparently this is how she conceived *The Italian*) very cautiously. And yet, he decided to retain those vestiges of the supernatural which in approaching the narrative material provided. In contrast to the vivid "diablerie" of *The Monk* and even to the terrifying mysteries of her *Udolpho*, *The Italian* is almost devoid of suggestions of supernatural agency and presence. This novel depends for its Gothicness almost exclusively on a skilful handling of the "monkish" element, i.e. Schedoni and the

40 In his biography of actor and manager John Philip Kemble (Memoirs of the Life of John Philip Kemble, Esq. Including a History of the Stage, from the Time of Garrick to the Present Period, published in London in 1825), Boaden discussed at some length the representation of the ghost in his play. See also Ranger's discussion of this scene in his "Terror and Pity reign in every Breast," p. 76.

other monks who, in their different roles and guises, haunt the narrative. For the sake of a sentimental reconciliation, Boaden has transformed the Radcliffean prototype into a reformed villain who halfway through the play not only abjures his wickedness but even decides to help Ellena to have her paramour, Vivaldi (*IM*, end of Act II).

In *The Italian*, the mysterious figure of the monk who appears in the first chapter is not only cloaked in mystery but has an air of the supernatural about him. No wonder, thus, that Boaden leaps eagerly on this figure and uses it as a source of suspenseful supernaturalism. In the play, the monk appears twice, and his entrances correspond to those in the novel. By a well-established convention (originating in Walpole), the servants in Gothic fiction are proponents of the superstitious belief in the supernatural. In the play this role is performed by Paullo (Radcliffe's Paulo). In scene I/2, it is Paullo who is responsible for a supernatural representation of the mysterious monk: "His face seemed the spectre of a long fast – He glared upon me with eyes flaming in sockets a foot deep in his head, and the motion of his arm, the very wind of it laid me prostrate on the ground. – *(A bell tolls suddenly at a distance – the monk at the stage speaks.)* – Vivaldi, hark! the knell of death sounds heavily! All is accomplished. [*Exit*" (*IM* I/2; 11-12).[41] Even without an epilogue, the audience must have been conscious of the Shakespearean inspirations of this mode of the supernatural.

Another scene which features the mysterious monk and realises the Radcliffean method of the explained supernatural is found at Act III Sc. 2 (corresponding to chapter V in volume III of *The Italian*), where Vivaldi, imprisoned in the dungeons of the inquisition, is visited by that same mysterious monk. Predictably and despite the usual compression, Boaden used verbal echoing. "To-morrow night you will meet me in the chambers of death!" says the monk to Vivaldi in *The Italian*.[42] And Boaden makes his Monk say: "Yet we shall meet, and in the hall of death" (*IM* 61).

Finally, *Aurelio and Miranda*, Boaden's five-act adaptation of *The Monk* (Drury-Lane, Dec. 1798) is another example of the processes involved in the transference of the literary material of Gothic fiction onto the late eighteenth-century London stages. Typically, Boaden has chosen some spectacular episodes in the novel, i.e. those of specially high dramatic tension. Boaden's faithfulness to his source decreases, however, as the plot evolves; i.e. he gradually departs from the plot or major plots of *The Monk*. Starting with fairly close verbal echoes, paraphrases, and occasional direct borrowings, he eventually stages a denouement which annuls the theme of the source narrative.

The title of the play names a pair of central characters who are not found in *The Monk*. A closer inspection reveals that this was conceived as a thin disguise

41 The spelling, punctuation, and italics of the original have been retained.

42 Ann Radcliffe, The Italian or the Confessional of the Black Penitents (Oxford: Oxford University Press, 1998), p. 323.

that allowed Boaden to distance himself from the novel. From scene one the protagonists are easily recognisable as alliterative counterparts of Ambrosio and Matilda, respectively. Furthermore, some other characters known from Lewis's book have been reused by Boaden in nearly unchanged roles, especially Antonia and Agnes, although in Boaden the latter is basically the sole object of monastic cruelty.

Boaden adapts a number of dramatic episodes from the two major plots of *The Monk*. He is also at pains to make them intertwine, and attains this goal more successfully than Lewis by making his monk, Aurelio, actively prevent Agnes's demise at the hands of the cruel Prioress. In particular, Boaden stages the high-tension letter episode: Raymond planting a letter which was intended for Agnes and which contained the elopement plan, Aurelio intercepting it, and delivering Agnes into the hands of the Prioress. In borrowing this sequence and presenting it in a number of scenes, Boaden used his theatrical instinct, but then Lewis also displayed a keen sense of the dramatic when composing the original narrative, itself loosely based on *Measure for Measure*. One thing is certain; Boaden must have known that most of what comes later in the novel, *despite* its dramatic potential and *because* of its scandalous attractiveness, was not to be tampered with. Having to make do with only a morsel of the source, he did not have to put too much effort into his endeavour, as numerous paraphrases, more or less closely echoing the text of the novel, show.

This is also true of the other major plot, that involving Aurelio and Miranda. In a succession of scenes, intertwining with those of the parallel Agnes plot, Boaden stages first the exposure of Miranda and then the gradual manner in which Aurelio succumbs to carnal desires. Despite numerous verbal parallels, Boaden treads here more and more cautiously as the story heads in the direction that it took in the source. While in Lewis the episode ends in an immoderate release of the now unbridled carnal passions, Boaden suddenly veers off this track and transforms his Miranda into a paragon of maidenly virtue. This largely explains the dissatisfaction of many in the audience, including Lewis.

Ultimately Boaden gives an elaborate scene of recognition and reconciliation, preceded by a rather forced moral reform of his monk.[43] His denouement, staged in the vaults of a monastery, due to many earlier departures from the adapted material, differs in meaning from the original narrative. Despite or even contradictory to the classic Gothic setting (on whose grisly interior Lewis dwells in his book with much relish) the scene is one of elaborate reconciliation. Released ("absolved") from the monastic vows (miraculously, thanks to his newly found identity), Aurelio can now enjoy Miranda (who displays none of Matilda's voluptuous abandon) as his

43 Lewis included a scene not unlike this one in The Castle Spectre, which once more shows Boaden in the role of an emulator. Chronologically, Boaden's Aurelio and Miranda came a whole year after The Castle Spectre (Evans, Gothic Drama, p. 241). The substantial difference is that Lewis' reconciliation scene features an avenging spectre, which of course makes it "more Gothic."

legitimate spouse. Aurelio actually rushes to Agnes's rescue, tormented by a sense of guilt; Antonia, instead of falling victim to Aurelio's unbridled lust, "kneels before [her] brother."

To sum up, Boaden rewrites the plot of the novel so as to achieve his pre-established goals, very different from those of Lewis: vindication of propriety and praise of "just indulgence" of "our passions." The main theme of *The Monk*, i.e. the moral corruption of Ambrosio, his headlong fall from sainthood into demoniacal sinfulness, is here merely hinted at as a remote possibility but never "shown," or realised mimetically. Miranda is at the same time an object of his "passion" and an instrument of his moral correction, his "monistress," as he puts it in the final speech of the drama. The scandalous elements of *The Monk*, the spectacular display of violation, persecution, supernatural agency, etc. – all the ingredients of "horror Gothic" have been all but entirely excised. Mere traces of the original horrors are retained in the verbal and diegetic (as opposed to what is actually presented upon the stage) domain of the portrayed world of the play. In IV/1, for instance, Raymond describes his dream, a descent down a subterranean passage and into a sepulchre, where he finds Agnes, dead in a tomb; suddenly this nightmare turns into a vision of a resurrection, featuring Agnes and her son rising from the tomb. In other words, the horrors or suggestions thereof are diegetically contained and eventually transformed into vehicles for hope-inspiring image and elevating sentiment.

The above examples show that Gothic on the stage was only possible as a peculiar cultural compromise. Private reading albeit confined to the Closet enjoyed more liberty than theatrical consumption of seemingly identical content. Being a sphere of public cultural transactions, theatre was considerably constrained, by the rules of decorum, moral decency, urbanity, taste, etc., which, oddly, made it a platform for the propagation of these Augustan standards. Boaden's dramatisations of the major hits of Gothic romance reflect the ambiguity involved in the endeavour: he sought to capitalise on the morally dubious and rapidly degenerating literary form. Yet, operating in the public domain of melodramatic and operatic theatre, he was not able to keep the promise that he was making. Paraphrasing the Porter's lines from *Macbeth* we may say that the cultural forces involved made Boaden equivocate; he provoked and unprovoked, he provoked the desire but took away the performance.[44] What attracted viewers (that they were readers of Radcliffe and Lewis was of course Boaden's premise) to the theatre was the desire to re-live and somehow verify (thanks to the different medium or mode) the experience they had while reading. Yet, if the theatre made this kind of promise, which seems to be the rationale behind the very idea of adaptation, then it was one that could not be fulfilled. Radcliffe and Lewis may have understood this. But

44 See Macbeth, II.iii.28, on drink being the great equivocator with lechery: "it provokes, and unprovokes; it provokes the desire, but it takes away the performance."

while Radcliffe seems to have despised the medium, Lewis eagerly exploited it, yet, like Boaden, did not attempt to represent on stage the horrors of his Gothic.

Boaden set out to adapt both the major modes of Gothic fiction, the terror and the horror variety, but seems to have sustained a double failure. For both these modes seem unadaptable as such, i.e. as they were originally conceived and executed in the novel form. On the one hand Boaden seeks to out-Gothicise Radcliffe's *The Romance of the Forest* and thus inevitably compromises her method; on the other, he tries to domesticate Lewis's demons and contain the scandalising luridness and diablerie of *The Monk*. In adapting Radcliffe, Boaden faces an artistic or technical problem; in adapting Lewis a moral one. Radcliffe's Gothic, immaculate as to the principles it preaches, is ideally tailored for solitary perusal and aesthetically inimical to the very idea of performance. Its "beauties" are dissipated in the compressed treatment which drama requires; in particular, the medium of the theatre does not tolerate the ambiguities and obscurities productive of narrative suspense. Lewis's Gothic, on the other hand, is scandalously spectacular or lurid but morally objectionable and therefore also unfit for public representation in the theatre.

Despite this unadaptability of both the Gothic modes he worked with, Boaden thought it worthwhile to give it a try and did not give up easily. Impelled by basically antagonistic forces, his adaptations are instructive in the very ambiguity of the authorial predicament that they allow us to inspect. And perhaps his adaptations are more than just so many failures. From a certain point of view, which apparently was that of Boaden, the stage was a platform, not for the final demise of Gothic, but of reconciliation, of adoption and of adaptation in a socio-political sense, that of the containment of an unruly mode of fiction, of making it acceptable and, to some extent at least – respectable. And yet, Boaden's Gothic exploits also tell a sad story, a story which makes "recycling" an adequate term for this type of treatment of existing literary "material." For no matter what input stuff the literary recycler chooses to lay his hands on, he principally treats it as waste, out of which eventually something new can and should be made, if only by the crude methods of crushing, pulling apart and soldering back together. The thus manufactured monster is perhaps the only real spectre which has ever haunted Gothic.

Bartosz Wójcik

Ridin' De Riddim. Sampling The S(Hit/Y)Stem. Benjamin Zephaniah as a Cultural Recycler

In a recent Guardian article Naomi Alderman wrote:

> I am eagerly anticipating the reaction to the Critical Studies in Television conference in Manchester next month, if only because it will afford me the always-thrilling opportunity to throw a newspaper across the room in disgust. If it is a slow news week, I confidently expect a flurry of snide pieces pointing out papers on "Xena Warrior Princess and Renegotiations of Place" ... and bemoaning the death of serious scholarship.
>
> I have never understood why there is such snobbery about studying modern pop culture. Does no one understand that some of today's pop culture will become high culture in a few decades' time? ... But of course, we can't have academics actually discussing something relevant to modern life, can we?[1]

The rationale behind my opening with such a quotation is simple: I am about to discuss the works of Benjamin Zephaniah – "possibly the most popular poet in England,"[2] a performance poet "that confronts you in a strangely populist fashion,"[3] a writer to whom Carolyn Cooper might have adverted when she diagnosed the maladies of reggae-based oral poetry: "The search for the exact word & rhythm is abandoned as the dubber [poet] settles for the automatic reflex of cliché."[4] His is poetry of mass participation, its journalistic themes[5] follow in Louise Bennett's footsteps, as Zephaniah "will use any means to communicate."[6] Being socially programmatic resulted in his being branded "a cutesy multiculturalist for soft-hearted weepies,"[7] and the user-friendliness of his works

1 Naomi Alderman. "It's shockingly funny...," *The Guardian*, 14 Aug. 2006: 15 Aug. 2006 <http:// film. guardian. co. uk /features/featurepages/0,,1844261,00.html>.
2 Bruce King, "*Propa Propaganda*" [Review], *World Literature Today*, vol. 71, no. 3 (1997), p. 595.
3 Gerry Hectic, "*Naked*" [Review]: 12 Apr. 2006: <www.fly.co.uk/fly/archives/ 2005/02/ benjamin_zephaniah _naked. html>.
4 Carolyn Cooper, "Word Unbroken by the Beat: The Performance Poetry of Jean Binta Breeze and Mikey Smith," in Anna Rutherford, ed., *From Commonwealth to Post-Colonial* (Amsterdam: Dangaroo Press, 1992), p. 199.
5 "[P]oems for NOW," reads the back cover of Zephaniah's *The Dread Affair*.
6 Bruce King, *The Internationalization of English Literature* (New York: Oxford University Press, 2004), p. 195.
7 King, *The Internationalization*, p. 197.

generated "controversy over the value of Zephaniah's poetry,"[8] which one critic "linked with the inherently oral nature of his art which can be partly explained by his late development as a reader and writer."[9] The present paper, however, does not aim to make any critical assessment of literariness of Zephaniah's poetry; neither does it centre on the artistic value of the poet's output. Therefore, rather than providing a deep-sea analysis of a sample of Zephaniah's output, I will skim stones over the surface of his extensive body of work with intent to accentuate the poet's literary strategies which involve musical techniques of recycling, sampling and remixing.

Driven by his politics of democratisation, which I have discussed elsewhere, [10] Zephaniah embarked on the task of broadcasting i.e. penning, performing and publishing accessible poetry, acting on the contention he later expressed in "Rapid Rapping (rant)":

> Long time agu before de book existed
> Poetry was oral and not playing mystic
> Poetry was something dat people understood
> Poetry was living in every neighbourhood
> Story telling was compelling listening, an entertaining
> Done without de ego trip an nu special training
> Found in many forms it was de oral tradition
> When government said quiet, poets said no submission.[11]

The intended consequence of utilizing creolized, if mildly, English and frequent application of Rastafarian imagery along with Dread, if curtailed, Talk was to dissociate the poet from any dialogue with English literature. His politics of initial en bloc contesting of the master narrative of the English literary canon is reminiscent of Langston Hughes' ("[m]elodramatic maybe, it seems to me now"[12]) gesture of dumping overboard his personal book collection, which he related in his 1940 autobiography *The Big Sea*.

In "Just Like David," Zephaniah's early 80s artistic manifesto, the biblical David is preferable to Dylan Thomas; Shakespeare is written off as a crowd-pleaser:

> Reading one the other day who spoke of under milk and wood,
> when I finished reading this it didn't learn me any good,

---

8 Eric Doumerc, "Benjamin Zephaniah: The Black British Griot," in Kadija Sesay, ed., *Write Black, Write British. From Colonial to Black British Literature* (Herford: Hansib, 2005), p. 195.
9 Doumerc, "Benjamin Zephaniah," p. 195.
10 Bartosz Wojcik, "Exposing the Great Baton's Underbelly. Benjamin Zephaniah's *Belly of De Beast* as a Politicised Concept Album." The article was presented at the Poetry and Politics Conference at the University of Stirling, July 2006.
11 Benjamin Zephaniah, "Rapid Rapping (rant)," *City Psalms* (Highgreen: Bloodaxe, 1992), p. 39.
12 Langston Hughes, *The Big Sea* (New York: Hill & Wang, 1993), p. 3.

reading one the other day who spoke of some midsummer dream,
what a silly waste of paper cashing the commercial scene.[13]

The disdain for the dominant literary culture will resurface in "Lesson Number Wan" from *School's Out*, his pupil-oriented collection:

English ideas nu work.
...
Slavery is not over
Philip Larkin means nothing to me [14]

However, it is not only the contents of curricula that Zephaniah targets. If he opts for an English ballad, it is to cast liberating anathemas, as in "Master Master." Here, in accordance with Rastafarian lexis, the *children* stand for generic Africans[15]:

If slave drivers be men of words,
We curse that poetry,
Its roots you'll find are so absurd,
Come children see. [16]

Ironically, within two decades Zephaniah skyrocketed to become a school book entry. Such is the nervous postcolonial condition of postmodern REAL-life diegesis.

The nineties brought qualified modification of Zephaniah's poetic modus operandi. The critical dialogue with Anglophone texts, including canonical poems and popular songs, that he partakes in resembles Linton Kwesi Johnson's acknowledgement of having "arrived at the position that Blacks don't live in isolation in England and that our fate is tied with other events that are taking place in other parts of the world."[17] Thus, Zephaniah engages himself with the body of English literature, which may be symbolized by a "passionate, poetic intercourse"[18] with Adrian Mitchell, whose "Criminal Justice for Crying Out Loud – A Rant" impregnated Zephaniah with a "sister poem"[19]: "Silence in Our Screams," a text on "miscarriages of justice."[20] Shakespeare is no longer a representative of the former "commercial scene," though Zephaniah persists in remaining one of those poets

13 Benjamin Zephaniah, "Just Like David," *The Dread Affair* (London: Arena, 1985), p. 32.
14 Benjamin Zephaniah, "Lesson Number Wan," *School's Out: Poems Not For School* (Edinburgh: AK Press, 1994), p. 3.
15 In BZ's "The President is Dead Again," the Africans are characteristically referred to as "men of words."
16 Benjamin Zephaniah, "Master, Master," *Propa Propaganda* (Newcastle: Bloodaxe, 1996), p. 59.
17 Billy Bob Hargus, "Linton Kwesi Johnson" [Interview]: 25 Nov. 2004 <http://bednorz.uni2.net/sbn/interv/lkj.htm>.
18 Benjamin Zephaniah, "Silence in Our Screams," *Propa Propaganda*, p. 33.
19 Zephaniah, "Silence," p. 33.
20 Zephaniah, "Silence," p. 32.

"who do it reggaematic,"[21] for whom poetry is of an indwelling nature. It is to its *rootsical* accompaniment that the speaker cycles in pursuit of his daily activities:

> Dis poetry goes wid me as I pedal me bike
> I've tried Shakespeare, Respect due dere
> But dis is de stuff I like.[22]

As the man pedals his bike, English words chase him and a marked verbal hunt takes place: physical cycling and literary recycling. Zephaniah's poetry interweaves oral and scribal modes; the poet, especially in his later works, creates Black Atlantic ethnoscapes, often grafting them onto canonical English templates. His undertakings involving textual recycling may be subsumed under two broad categories: replication and re-editing. Speaking of these, I shall refer to Zephaniah's usage of paradigmatic reggae and hip hop recycling-reliant techniques, which enable the writer, who once portrayed himself watching Aston Villa "with my Jamaican hand on my Ethiopian heart/The African heart deep in my Brummie chest,"[23] to confer with English and Black Atlantic traditions.

His poetic renderings, such as the one I am about to discuss, are not equivalents of mere musical covers in the Western understanding of the term. Although Zephaniah's emulation of the syntactic and prosodic qualities of Anglophone texts is reminiscent of, for instance, Kenneth Koch's "Variations on a Theme by William Carlos Williams," I am inclined to classify his poems as individualised reworkings of original literary compositions set in a Jamaican context. Reggae music hinges on the employment of drum and bass rhythm tracks, known as *riddims*, which provide templates for "countless recuts by raw hopefuls and seasoned talent alike."[24] The popularity of certain reggae/dancehall rhythms[25] gave rise to the emergence of the phenomenon of the so-called "riddim-driven albums" where a duplicate instrumental track is treated as a platform for various deejays and singers to showcase their unique vocal prowess. As an example, in 2005 Jamaican Junior Gong returned to prominence with a chart-topper called "Welcome to Jamrock"[26] whose foundation is the "World-A-Music" riddim dating back to the early eighties. The success of the Caribbean single sparked off a string of European follow-ups, including "Welcome to England" and "Witamy w Polsce."

21 Benjamin Zephaniah, "Rapid Rapping" in *idem*, *City Psalms* (Highgreen: Bloodaxe, 1992), p. 39.

22 Benjamin Zephaniah, "Dis Poetry" in *idem*, *City Psalms*, p. 12.

23 Benjamin Zephaniah, "Knowing Me," *Too Black, Too Strong* (Highgreen: Bloodaxe, 2001), p. 63.

24 David Katz, *Solid Foundation. An Oral History of Reggae* (London: Bloomsbury, 2003), p. 344.

25 According to an online database (www.riddimbase.net; retrieved on 9 September, 2006), Sleng Teng (1985), the first ever digital riddim, has been rendered at least 240 times; Real Rock, a classic 1970s riddim, has been reused over 350 times.

26 Jamrock = Jamaica, an island that remains both a precious stone and a country rocked by pitfalls.

Since the process of colonization changes the sovereign and the subaltern, a cultural transference occurs. Hence, Larkin's much-anthologized "This Be The Verse" is, to paraphrase Louise Bennett, "colonized in reverse" and treated as a reggae riddim by Zephaniah. His "This Be The Worst" is a riddim-driven *version* of Larkin's modern classic, much as Jamaican U Roy's "Natty Rebel" is a *version* constructed out of the rhythms and vocal harmonies of "Soul Rebel," initially conceived and performed by the Wailers:

They fuck you up, those lords and priests.
They really mean to, and they do.
They fill themselves at highbrow feasts
And only leave the crumbs for you.

But they were fucked up long ago
By tyrants who wore silly gowns,
Who made up what they didn't know
And gave the masses hand-me-downs.

The rich give misery to the poor.
It deepens as they hoard their wealth.
They'll be fucked up for ever more,
So just start thinking for yourself. [27]

Similar transference from one medium to another is materialized in "One Day in Babylon," which is also an instance of *riddim*-like replication, although this time Zephaniah – instead of utilizing all the prosodic and syntactic patterns of Emily Dickinson's "Because I could not stop for Death" – resolves to mirror just one couplet. Compare:

Because I could not stop for Death-
He kindly stopped for me - [28]

Because I told him I need friends
He would not go away [29]

Furthermore, "One Day in Babylon" doubles as an emulation of Emily Dickinson's poetic style. Thus, Zephaniah *rides de riddims* of easily identifiable poems, staple literary works included in school curricula, such as his "What If" in the mould of Kipling's "If." Zephaniah's *reggaematic version* does not travesty the original penned by the author of "Gunga Din" but retains its Kiplingesque advocacy of moderation and self-restraint, modernizing the content to be pertinent to a Cruel-Britannia context:

---

27 Benjamin Zephaniah, "This Be The Worst," in *idem*, *Too Black, Too Strong*, p. 30.
28 Richard Ellmann and Robert O'Clair, eds., *Modern Poems* (New York: Norton, 1989), p. 25.
29 Benjamin Zephaniah, "One Day in Babylon," *idem*, *Propa Propaganda*, p. 36.

> If you can make one heap of all your savings
> And risk buying a small house and a plot
> ...
> If you can speak the truth to common people
> Or walk with Kings and Queens and live no lie[30]

By means of such echoes, Zephaniah welcomes identification on the part of the audience, already acquainted with these school, as opposed to scholarly, texts. Estrangement does not take place since rhythmical and lexical building blocks need no expository notes. Rarely does he also include footnotes to credit the hip hop-like textual samples that he often avails himself of.

His rendition of "What a Wonderful World," Louis Armstrong's crossover all-time classic, is, however, supplied with a disclaimer in which Zephaniah praises the late Satchmo but elects to "walk the same road and see things from a different point of view."[31] His is a "Terrible World" populated by abusive "pimps and priests/Well interfused"[32] and gradually depopulated as "mass graves [are] made/from a football pitch."[33] At this point, it is worth remarking that during performance, which "itself thus becomes a privileged reading of the text,"[34] Zephaniah models his delivery on Armstrong's bellowing voice and timbre. Once again, "What a Wonderful World" may be considered a riddim.

Surprisingly, the Bob Marley-inspired "I Neva Shot De Sheriff" follows a different scenario. It is a dramatic monologue delivered posthumously by one of the Black males "professionally done"[35] by police officers: "Word on TV is,/Twaz a precautionary measure."[36] As the victims "are still waiting/for [...] public enquiry,"[37] the poem reverses the roles played by the characters of Marley's "I Shot the Sheriff" and, as is Zephaniah's wont, accuses the *shitstem*:

> I shot the sheriff, but I swear it was in self defense
> ...
> I, I shot the sheriff, but I didn't shoot no deputy [38]
> De sheriff shot I an me
> An den he shot me deputy.[39]

30 Benjamin Zephaniah, "What If," in *idem*, *Too Black, Too Strong*, p. 17.
31 Benjamin Zephaniah, "Terrible World," in *idem*, *Propa Propaganda*, p. 13.
32 Zephaniah, *Propa Propaganda*, p. 13.
33 Zephaniah, *Propa Propaganda*, p. 13.
34 Cooper, "Word Unbroken by the Beat," p. 197.
35 Benjamin Zephaniah, "I Neva Shot De Sheriff," in *idem*, *Too Black, Too Strong*, p. 41.
36 Zephaniah, *Too Black, Too Strong*, p. 41.
37 Zephaniah, *Too Black, Too Strong*, p. 44.
38 Bob Marley, "I Shot the Sheriff": 23 May 2006
<http://www.bobmarley.com/songs/songs.cgi?sheriff>
39 Zephaniah. "I Neva Shot De Sheriff,"p. 44.

A reader prone to espying cross-references will notice that Zephaniah's "De Rong Song" refracts the optimism of Bobby McFerrin's "Don't Worry, Be Happy." By this act of transference, Zephaniah rebels at the thought of being reduced to a Black entertainer. In his 1980s poem, "The day dat I met Lady Di," the People's Princess confused him with Lenny Bruce and Eddy Grant,[40] filing him under Anonymous Other as a result. The title's obvious reference to Frank O'Hara's 1959 "The Day Lady Died" measures the American's reverence for Billie Holiday against Zephaniah's and Lady Diana Spencer's mutual irrelevance.

Such referentiality characterizes both reggae music [41] and reggae-engendered oral poetry. In her "Slam Poem,"[42] Jamaican Jean "Binta" Breeze recycles the lyrics of Beanie Man's hit single "Who Am I" to create a palimpsest-like commentary on the dancehall performer's Alpha male hedonism. Quite often, as in the case of Zephaniah's "I Neva Shot De Sheriff," referential poems may be viewed as unofficial sequels or fan fiction. Jean "Binta" Breeze's "The Wife of Bath Speaks in Brixton Market"[43] and Patience Agbabi's "The Wife of Bafa"[44] provide prime examples; both are contemporary renditions of Chaucer's tale, and both serve as alternative histories, updates on the mediaeval character.

Benjamin Zephaniah's "I Have a Scheme" is a poem which once more brings the reader back to familiar surroundings. Contrary to Bruce King's assertion, to my mind it is not solely a "comic imitation"[45] but a spatial and temporal transposition, an anglified version of Martin Luther King's 1963 "I Have a Dream" speech. This is what Tony Blair's Cool Britannia might have looked like as the micronarrative of "I Have a Scheme" questions the ubiquitous macronarratives of nationalism. Zephaniah localizes the momentous speech, replacing the original Negro spiritual ("Free At Last") with a gospel of his own: "[...] that great old English spiritual, */Here we go, Here we go, Here we go*"[46] i.e. the prototypical football chant and/or a strikers' rallying cry. "Life, Liberty and the pursuit of Happiness" that Dr. King (and the signers of the Declaration of Independence) espoused evolve into a "pursuit of a truly British way of life." Such modern mores question monolithic social/gender/ethnic/class etc. roles, to which the hospitality of pluralism is the answer:

---

40 In "If I woz a tap notch poet" LKJ referred to being pigeonholed as "a black Lance Percival in reverse."

41 See, for instance, Burning Spear's "As It Is" (*Calling Rastafari*, RAS Records 1999) – a reworking of his 1970s "Slavery Days."

42 Jean Binta Breeze, "Slam Poem," *The Arrival of Brighteye and Other Poems* (Newcastle: Bloodaxe, 2000), pp. 42-46.

43 Jean "Binta" Breeze, "The Wife of Bath Speaks in Brixton Market," *The Arrival of Brighteye and Other Poems* (Newcastle: Bloodaxe, 2000), pp. 62-64.

44 Patience Agbabi, "The Wife of Bafa," *Transformatrix* (Edinburgh: Payback Press, 2000), pp. 28-9.

45 Bruce King, p. 595.

46 Benjamin Zephaniah, "I Have a Scheme," in *idem*, *Propa Propaganda*, p. 10.

And in that time
Afro-Caribbean and Asian youth
Will spend big money on English takeaways
And all police officers will be armed
With a dumplin ...
Immigration officers will just check that you are all right
And all black people will speak Welsh …
I see thousands of muscular black men on Hampstead Heath
walking their poodles ...[47]

Apart from creating modified copies (sundry versions), Zephaniah recycles texts by dint of re-editing. His method of recontextualization resembles hip hop sampling, namely the cut-and-paste technique. Zephaniah's "Room For Rent" contains a replayed/rewritten sample of Michael Smith's "Me Cyaan Believe It:" "Room dem a rent/me apply widin"[48], which – as analysed by Carolyn Cooper[49] – is Smith's evocation of a children's ring game popular in the Caribbean. Zephaniah's adapted citation ("Room for rent, apply within"[50]) provides a formal structure for the poem, which thematically does not diverge from the original. Inside the vacant room, in lieu of Smith's "cockroach rat an scorpion," there rules a prejudiced landlord. Outside, there are, respectively, hostile neighbours and equally hostile fellow Blacks.

As popular music intertextuality is reintroduced with a vengeance by the refrain of "Belly of De Beast," Black Atlantic sampling becomes a device and a mission statement. The refrain of the poem/song recalls "Jailhouse Rock" – the 1957 movie that catapulted Elvis Presley to fame and locked him in the house of ill fame. It was then that he allegedly expressed racist sentiments which – coupled with his Mississippi descent (and the Memphis mansion he acquired) – fuelled the outcry of Black America over "a white performer whose financial success rested upon the songs and styles of black artists historically excluded from the popular music marketplace."[51] The allusive if elementary refrain, namely: "Dis is not De Jail House Rock,"[52] addresses the issue of commodification (tackled and namechecked, among others, by Public Enemy's "Fight the Power," Living Colour's "Elvis is Dead," Mos Def's "Rock n Roll," Mutabaruka's "Thieving Legacy" and Amiri Baraka's "In The Funk World"), casts doubts on Presley's non-racist behaviour and juxtaposes the misery of captivity inside the belly of the

47 Zephaniah, *Propa Propaganda*, p. 10.
48 Michael Smith, "Me Cyaan Believe It," *It A Come* (London: Race Today, 1986), p. 13.
49 Cooper, p. 210.
50 Benjamin Zephaniah, "Room For Rent," in *idem*, *School's Out* (Edinburgh: AK Press, 1994), p. 23.
51 Michael T. Bertrand, *Race, Rock and Elvis* (Illinois: University of Illinois Press, 2000), p. 26.
52 Benjamin Zephaniah, "Belly of De Beast," in *idem*, *Propa Propaganda*, p. 15.

Babylonian beast with the prison merriment depicted in the original "Jailhouse Rock" song:[53]

> The warden threw a party in the country jail ...
> Shifty Henry said to Bugs: For heaven's sake,
> No one's looking, now's our chance to make a break.
> Bugsy turned to Shifty and he said: Nix, nix,
> I wanna stick around a while and get my kicks.[54]

A further dethronement of Presley (though the song "Blue Suede Shoes" was written by Carl Perkins) and the whitened, neutered rock for which he stands is carried out by the expedient of yet another negation, which recurs in "Nu Blue Suede Shoes":[55]

> Me nu hav nu blue suede shoes
> But me really want fe rock.
> ...
> Der is music in me heart
> So please will you excuse me
> If my shoes are not so smart,
> Please will you excuse me
> While I tear de state apart.[56]

The titles of popular songs sampled and cited by Zephaniah are either tampered with, as in the cases of Presley and McFerrin, or treasured, as in the poem "Dancing the Tradition." Here the tradition is defined by Bob Marley's heritage, the titles of whose songs, spanning Marley's entire recording career (from the early "Get up, Stand up" to the posthumously released "Iron Lion Zion") are incorporated into the semantic body of the poem – a textual practice reminiscent of Mutabaruka's "Wailin." Correspondingly, the snippets of texts extracted from the lyrics of reggae songs, recycled and fashioned into the lines of Zephaniah's poetry, function in the same way as the sample from Mutabaruka's "Hati" which Zephaniah turned into the title of his 2001 collection, namely *Too Black, Too Strong*. The assorted citations are references to Jamaican/Black Atlantic, culture(s), and reminders of its/their re-cyclicity. For instance, the first stanza of Zephaniah's "Uptown Downtown" opens with a quotation of parallel verses of U Roy's song "On the Beach,"[57] which in turn is based on the rhythm track and vocal harmony

53 According to Wikipedia, "Jailhouse rock (JHR) is an indigenous African American fighting art that has its origins in the 17th and 18th centuries, when slaves were first institutionalised and needed to defend themselves," 4 Sep. 2006: <http://en.wikipedia.org/wiki/Jailhouse_rock>.

54 http://en.wikipedia.org/wiki/Jailhouse_Rock/song (retrieved on 7 July 2006).

55 Jamaican Mutabaruka's "Killin" includes tell-tale "blood suede shoes." See: Mutabaruka, *The Next Poems (1980s-2000s)* (Kingston: Paul Issa Publication, 2005), p. 44.

56 Benjamin Zephaniah, "Nu Blue Suede Shoes," in *Too Black, Too Strong*, p. 79.

57 It is a common practice among reggae artists to quote fellow musician's lyrics. For instance,

recorded by the Paragons, a Jamaican rocksteady group. Zephaniah's "Naked," a poem appreciated by Kwame Dawes, who generally lambasts Zephaniah's cross-fertilizing endeavours ("if he borrows from a poetic tradition, it is in a thinly read British verse tradition that he never stretches or challenges"[58]), includes samples among which are modified quotes from The Rolling Stones' "It's Only Rock'n'Roll" ("I know it's only Hip Hop Rock but I like it") and Peter Tosh's "I'm the Toughest" ("I have come to realise that what you can do for me/I can do much better for me"). The poem assimilates also the elements of Psalm 133 (133:1), popularized by Bob Marley in his "Africa Unite," written when the war atrocities in Angola and Mozambique escalated:

> Behold, how good and pleasant it is
> when brothers dwell in unity[59]
>
> How good and pleasant it would be
> Before God and man, yeah
> To see the unification of all Africa, yeah[60]
>
> Behold how good & how pleasant
> It is for revolutionaries to dwell together
> in the house of the Lord[61]

Much as Zephaniah's poetry is democratised, the tradition of a West African griot that he draws on is europeanized. As an "Equal Opportunities poet"[62] he stresses, in a postmodern fashion, the plurality of comprehensions and cultural fluidity/indeterminancy ("Me in Africa a Squatter/Me at home a Refugee"[63]), while by and large emphasizing qualities like objectivity, accentuating the notion of truth, typically in synch with roots reggae's reliance on transcendental signifiers e.g. God/Jah, realism. Bogged down in the Rushdie-defined "bogy of Authenticity," if you prefer.

The seeming clash, as construed by some commentators, [64] between Zephaniah's identities – unsparing poetic persona and amiable public/stage presence, may lead one to an act of untimely critical defusing of his social and incendiary poetry. He might, to evoke Mutabaruka's late 1970s poem, have turned into one of the revolutionary poets who "ave become entertainers," [65] yet

---

Black Uhuru's 1981 "Chill Out" cites the words of Dillinger's 1976 "Cokane in My Brain."
58 Kwame Dawes, "Too Black, Too Strong" [Review]: *World Literature Today* 76/ no. 2 (2002), p. 159.
59 *The Holy Bible English Standard Version*: 3 Jun. 2006 <http://www.gnpcb.org/esv>.
60 Bob Marley, "Africa Unite": 7 May 2006 <www.bobmarley.com/song/song.cgi?africaunite>.
61 Benjamin Zephaniah, "Naked," in *idem*, *Too Black, Too Strong*, p. 49.
62 Benjamin Zephaniah, "I Have a Scheme," in *idem*, *Propa Propaganda*, p. 10.
63 Benjamin Zephaniah, "Questions," in *idem*, *City Psalms*, p. 57.
64 Dawes, "Too Black, Too Strong" [Review], pp. 159-160.
65 Mutabaruka, "Revolutionary Poets," *The First Poems (1970-1979)* (Kingston: Paul Issa, 2005),

Zephaniah's work remains "a hybrid, creolized product of the meeting of two cultures ..., steeped in the very British tradition of doggerel and nonsense while paying homage to the Caribbean oral tradition in its various guises."[66] Studying his poetic output in a diachronic fashion, one notices that to him a poem, particularly a non-scribal one, as he prefers to be dubbed an oral poet,[67] is a work in process, refashioned according to the mode of presentation. "Empire," a danceable protest song that Zephaniah recorded with Sinead O'Connor, remixes his previously published "The Curse of Count Empire" and rehashes its reggae topos of a vampire, of a vampiric, beastly system. Similarly, the imagery of 2000,s "Homesick" bears the hallmarks of "Dis Policeman Keeps On Kicking Me to Death," his 1980s prototypical protest poem.

Zephaniah, a poet whose individual talent makes him not oblivious to past traditions and present trends, takes heed of the baffling symbiosis of the karaoke of voices, both recycled and alluded to, rooted in his work. In "Translate" he voices his concern over insurmountable communication barriers, intermittently built and broken by cultural recycling:

Who will translate
dis stuff.
Who can decipher
De dread chant
Dat cum fram

De body
An soul
Dubwise?
...
Sometimes I wanda

Who will translate
Dis
Fe de inglish?[68]

To conclude, it seems that once again it took yet another foreigner, i.e. my humble self, to answer Zephaniah's question and subsequently tackle the task of rephrasing his re-cyclicity. I can only hope that my argument has not been lost in the quagmire of translation.

---

p. 56.
66 Doumerc, "Benjamin Zephaniah," p. 195.
67 Benjamin Zephaniah, "Truth": 11 Mar. 2006 <http://www.benjaminzephaniah.com/truth.html>.
68 Benjamin Zephaniah, "Translate," in *idem*, *Too Black, Too Strong*, p. 83.

PART IV

RECYCLING THEORY

Anna Chromik-Krzykawska

Between Use and Refuse: Reclaiming the Abject into Culture

Re-cycling means bringing something back into a cycle. The recycling symbol with three arrows arranged in the form of an unending loop suggest a closed, finite circuit. Through its continuous motion it creates a sphere of constant tensions and dynamic interactions, rendering everything which is beyond it marginal. The marginal is not drawn into the dynamic current of movement, it is not engaged in any activity, it does not create energy, it is passive and useless – it is waste. The finite circuit does not acknowledge any gaps. The cycle seen as a certain procedure is a system following its patterns. If one of the subsequent phases is interrupted, the whole process might be disrupted. The dynamic tensions within the cycle guarantee balance and self-integrity.

The cycle is also a circle, the enlightened geometrical sphere standing for harmony, order and ratio. As such the cycle epitomizes the great modern narratives: economic utility, rationalism, purposefulness, wastelessness: values greatly esteemed by civilization. The cycle is then a sphere of regulation and meaning – a perfect embodiment of principles promoted by modern narratives. What is beyond this demarcated sphere is then "what disturbs identity, system, order" and "[w]hat does not respect borders, positions, rules,"[1] that is, what Julia Kristeva defines as abject. The abject, according to Kristeva, is an alien in the self, excreta, what disrupts the integrity of an "own and clean self."[2] Separation from the abject marks an entrance into the world of culture and a disconnection from non-differentiated pre-discursive chaos. Mary Douglas in her classic study of the subject shows that impurity can be analyzed as "that which departs from symbolic order"[3] and thus serves to delineate the borders of the symbolic and normalize the "self" and the "social" mediation.

Quite an obvious statement that "[p]eople get rid of excess"[4] made by social historian Susan Strasser might serve as a starting point for a deeper and more fundamental analysis. Trash, according to Strasser, is created by sorting and classification, processes which have a spatial dimension,[5] because "marginal categories get stored in marginal spaces."[6] The division between the central space

1 Julia Kristeva, Powers of Horror: An Essay on Abjection, trans. Leon S. Roudiez (New York: Columbia University Press, 1982), p. 4.
2 Kristeva, Powers of Horror, p. 53.
3 Kristeva, Powers of Horror, p. 91.
4 Susan Strasser, Waste and Want. A Social History of Trash (New York: Owl Books – Henry Hold and Company, 1999), p. 4.
5 Strasser, Waste and Want, pp. 5-6.
6 Strasser, Waste and Want, p. 6.

and the marginal space seems crucial to the study of cultural dimensions of pollution. The semiotics of boundary maintenance [7] seems fundamental for Douglas, who focuses on borders as delineating the sphere of order from that of expulsion. In her analysis of abjection, Kristeva also highlights the importance of demarcation, claiming that "separating, not touching, dividing, washing" are reactions to "the abolishment of limits and differences."[8] Orderliness is then a projection of the most desired state of certainty and regularity, a guarantee of a stable self; and consequently, various practices of cleansing and purifying are aimed at defending the integrity of the "clean and proper self."

Due to the importance of the demarcating qualities of the abject / *propre* distinction, the polluting factor is often essential to the constitution and well-being of the clean center. William A. Cohen writes about the constructive function of dirt, pointing to the importance of "what is marked as dirty, disreputable, or excluded from official culture" as "crucial to culture's self-constitution."[9] Defining the border between edible / inedible, pure / impure, and useful / useless serves to draw the limits of subject and object – whether it is an individual separation from bodily wastes, removing trash beyond the borders of the household, or cleansing cities of corpses and social impurities which are moved to some out-of-town cemeteries, infirmaries, or slums. Such efforts to eliminate dirt (or actually what is perceived as dirt in a given place and time) epitomize what Douglas describes as imposing system on "an inherently untidy experience" [10] and "not a negative moment, but a positive effort to organize the environment."[11] Rodolphe el-Khoury even claims that the organization of modern urban space with its institutions and social functions springs from the increasing significance of the hygienic / olfactory factors.[12] Division, separation and compartmentalization of modern flats, beds, or tombs – these all coincide with the reactions to the abject as described by Kristeva.

Cleansing is an exorcism – it neutralizes the transgressive and the marginal, deprives it of its disruptive qualities and renders it harmless, so that it can be safely reclaimed by the discourse and redrawn into the cycle of purposefulness. Purification does not always concern actual cleaning procedures – very often it takes place through rationalizing waste, endowing it with sense and meaning, and thus appropriating it back into the cycle – that is, recycling it. Cohen claims that

---

7 The term is used by Adeline Masquelier in her introduction to Dirt, Undress and Difference: Critical Perspectives on the Body's Surface (Bloomington and Indianapolis: Indiana University Press, 2005), p. 7.

8 Kristeva, Powers of Horror, pp. 159-160.

9 William A. Cohen, "Introduction," in William A. Cohen and Ryan Johnson, eds., Filth: Dirt, Disgust and Modern Life (Minneapolis and London: University of Minnesota Press, 2005), p. xvi.

10 Mary Douglas, Purity and Danger. An Analysis of Concepts of Pollution and Taboo (London and New York: Routledge, 2002), p. 5.

11 Douglas, Purity and Danger, p. 2.

12 See Rodolphe el-Khoury, "Introduction," in Dominique Laporte, History of Shit, trans. Rodolphe el-Khoury (Cambridge and London: The MIT Press, 2000), p. xi.

even the filthiest of substances have been taken into account in "utopian fantasies of reuse and regeneration."[13] He makes, however, a distinction between two categories of marginal entities that he designates "polluting" and "reusable." The former, he claims, is contaminating and therefore must be rejected or denied, whereas the latter embraces all things refused but potentially productive.[14] Such a division reverberates with Georges Bataille's discussion of the two ways in which the homogenous world of work deals with its waste. According to Bataille, knowledge, which is an integral part of the rationally disciplined world of work and reason, embodies the discursive procedures of the world of work: cognition is a necessary stage of knowledge, and knowledge is necessary to survive. But knowledge, like any other process, produces its offal. The excreta of knowledge might undermine the discourse: they open our eyes to the existence of heterogeneity, which is beyond the homogenous world of work. Therefore, as Bataille argues in his work on Marquis de Sade, the offal or the "foreign bodies" become a subject of two mutually exclusive processes – either of expulsion beyond the system, or of appropriation into the system.[15]

Although the examples that follow indicate that this distinction is valid in relation to the discourse of modernity, they also make clear that in a narrative of usefulness nothing is inherently rejected or denied, but abjection is a necessary stage followed by the system's attempt to re-absorb the previously marginalized entities. Such understanding of abjection corresponds to Mary Douglas's argument that dirt is simply "matter out of place,"[16] so nothing is actually dirty "in itself," or, as Susan Strasser argues, "[n]othing is inherently trash."[17] Dirt and trash are then to be considered labels attached to entities that fell out of the circulation, but their marginal existence beyond the cycle is only temporal. Such a view seems to correspond to Michael Thompson's theory of revaluation of waste objects. According to Thompson's "rubbish theory," objects are first devalued, that is, moved to the rubbish zone, when they have no value at all – that would be an equivalent of Cohen's "polluting" category. Then comes the process of revaluation, which seems to correspond to Cohen's idea of reusability.[18] To put it in other words, what is considered dirty / useless / out of place by a modern discourse is first marginalized, denied and excluded, but then there is a strong imperative to reclaim it into the system. The process of redrawing the marginalized abject back

---

13 Cohen, "Introduction," in Cohen and Johnson, eds., Filth, p. xii.
14 Cohen, "Introduction," p. x.
15 Georges Bataille, "La valeur d'usage de D.A.F de Sade, " in Oeuvres complètes II (Paris: Galimard, 1970), p. 59.
16 Douglas, Purity and Danger, p. 44.
17 Strasser, Waste and Want, p. 5.
18 Michael Thompson, Rubbish Theory: The Creation and Destruction of Value (Oxford: Oxford University Press, 1979), pp. 25-26. Quoted in Cohen, "Introduction," p. xv.

into the centre – that is, recycling it, bringing it back into the cycle – requires, however, certain exorcising procedures.

The procedure of reclaiming the abject into the sphere of order can be traced on the level of both the individual body and the social body which is the target of some normalizing practices that, according to Michel Foucault, shape, train and manipulate it through the regulations of army, school and hospital so that it becomes what he calls a "docile body" – a body that "may be subjected, used, transformed and improved."[19] The examples discussed below illustrate the working of such regulating procedures through, first, normalizing the abject body innards in the so called anatomical Renaissance, and second, exorcising the vertical city space in nineteenth-century sewage treatment.

Jonathan Sawday in his book about dissection in Renaissance culture designates the late sixteenth-century as the beginning of what he calls "the anatomical Renaissance."[20] It is not a coincidence that the anatomical Renaissance overlaps with the development of the normalizing discourse of modernity. The Cartesian body-machine is, as Foucault writes, a "body that is manipulated, shaped, trained, which obeys, responds, becomes skilful ...," [21] and also a perfect embodiment of the seventeenth century obsession with objectification, preventability and delineation. Such a body presents itself as more secure and predictable than the unknown land of the "natural" body, as "[t]the mechanical body dispelled doubt, uncertainty and indecision."[22] The omnipresent imperatives of instrumentation and artifice became the most common indicators of regulating bodily practices: "the body as a mechanism was now itself subject to mechanism, a technique, a field of productive labour which relied on ingenious inventions and instrumentation."[23] The mechanization of the body serves as its domestication, its absolution. The docile machine-body, unlike the unruly "natural" body, becomes fully acceptable. Abjectification of the "natural" body in the early modern era is linked with the discovery of the body's inner capacity. The body becomes a vast abyss with its "gloomy vaults, unsounded wells, panting furnaces, lakes of blood and urine."[24] This should be viewed in the context of the Renaissance depiction of a confrontation between interior and exterior as a conflict between the "grotesque" and the "classical."[25] What is inside the body is then associated with "low" culture, and to be reclaimed by the discourse must be refined, that is re-cycled, through the process of discursive externalisation and formalisation.

---

19 Michel Foucault, Discipline and Punish and the Birth of the Prison (New York: Vintage Books, 1995), p. 136.

20 Jonathan Sawday, The Body Emblazoned. Dissection and the Human Body in Renaissance Culture (London and New York: Routledge, 1996), p. 4

21 Foucault, Discipline and Punish, p. 136

22 Sawday, The Body Emblazoned, p. 32.

23 Sawday, The Body Emblazoned, p. 33.

24 Sawday, The Body Emblazoned, p. 19.

25 Sawday, The Body Emblazoned, p. 19.

The desire to externalize, the affirmation of the outside, might seem to run counter to the sixteenth- and seventeenth-centuries' exploration of the human body and the interest in anatomy, which actually looks "inside" and focuses on the internal. It must, however, be remembered that the aim of the anatomist is nothing but externalization, the flattening and delimitation of the vast and disturbing depths of the inside. The process of externalization is clearly visible in the Renaissance curbing of the undomesticated interior by mapping it in the anatomical atlas. The early modern body was perceived as a territory, an "undiscovered country, a location of which demanded from its explorers skills which seemed analogous to those displayed by the heroic voyagers across the terrestrial globe."[26] This period was an era of great discoveries, also within the body which was seen as *terra incognita*. The imagery of the colonization of the body was initiated, according to Jonathan Sawday, by the publication of *De Humani Corporis Fabrica* by Vesalius in 1543.[27] In their bold and dangerous explorations of the bodily interiors, the Renaissance anatomists were like "heroic voyagers across the terrestrial globe,"[28] acquiring the privileged status of mediators between the disturbing interior available only to the "disciplined eye" and the ordered language of science and control. What is more, the colonial process of discovering, naming and classifying wild bodily spaces is nothing but the flattening of this multidimensional immense depth and the reducing of it to a piece of paper in an atlas of anatomy, restraining and reclaiming the unruly into the scientific discourse. The discoverers even dot their names, like names of places on the map, over the body (for instance Eustachius mapping the terrain of the ear, Fallopius the female reproductive organs, or Serventus the pulmonary transit of blood).[29] The pre-discursive chaos is thus curbed and drawn into the cultural system of language, paper, schemes, surfaces, contours and external lines. The mapped inside is controlled by the cartographer's disciplined eye and ceases to be vast, threatening, and *unheimlich*.

The institution of the anatomy theatre also neutralizes the horror of a journey to the bodily interior by celebrating the process of autopsy and highlighting the formal aspect of this peculiar "ritualistic drama."[30] Allotment of seats in the auditorium according to social rank, music played during the performance, the procession heralding the entrance of anatomists to the *locus anatomicus*, the kind of language they used and the didactic function of the demonstration[31] seem to have compensated for the boldness of the anatomists / explorers who dared to venture into prohibited and sinister territories. Focusing on the ceremony and outer form of the anatomical performance is a perfect example of the Renaissance

---

26 Sawday, The Body Emblazoned, p. 23.
27 Sawday, The Body Emblazoned, p. 23.
28 Sawday, The Body Emblazoned, p. 23.
29 Sawday, The Body Emblazoned, p. 23.
30 The term is used by Sawday, The Body Emblazoned, p. 75.
31 For more information about anatomical theatres, see Sawday, The Body Emblazoned, pp. 73-75.

aspiration to externalize the internal and set a clear border on the amorphous. Bringing the hidden and uncanny gulfs of the human entrails to light performs a similar function to that of the atlas of anatomy: it domesticates creepy areas, reduces them to harmless theatre props, and thus reclaims them into the cultural discourse of usefulness.

Just as the vertical space of the body was seen as abject in the 16th century, the 19th century vertical space of the city inspired, according to David L. Pike, a mixture of fascination and repulsion. Unlike the urban subway-underground with its manageable and mapped system of geometrical tunnels, which was a modern epitome of progress and civilization, the sewer "remained closely associated with mythic and religious traditions of the epochal, life-altering trip to the underworld."[32] Pike describes two different reactions to the city's bowels: that of disposal, characteristic of London, and that of incorporation, attributed to Paris. Whereas the London model focused on flushing filth out of sight and repressing it, the Parisian strategy was that of naturalization aimed at containing, recycling and incorporating, often with a profit.[33]

A striking example of sewage-reclaiming utopia is to be found in Victor Hugo's *Les Miserables*. Just to quote a short fragment: "If our gold is filth, on the other hand, our filth is gold. ... All the human and animal manure which the world loses, restored to the land instead of being thrown into the water would suffice to nourish the world."[34] Cohen claims that although Hugo's reflections are embedded in a discourse on economic utility and organic cycles, they are also underpinned with a religious narrative of deep trust in the purposefulness of divine creation.[35] The modern narrative of wastelessness is marked by a sense of meaningfulness, in which the Great Clockmaker looks after the efficiency of divine mechanisms whose workings only need to be revealed under the layers of cultural prejudice and irrational disgust. Hugo's ideas were not pure fantasies – according to David Pike, his practical directions urged the recycling of Parisian sewage as fertilizer only four years after the publication of *Les Miserables*.[36] Although the drainage system could be seen as representing the "rational control of an archaic underground past,"[37] it still retained, as Pike argues, its symbolic significance of primitive, organic, irrational reverse of a lucid and controllable modern city space.[38] The descriptions of sewers echo with the imagery of the Lacanian Real, Kristeva's abject and Bataille's heterogeneity: they evoke regressive and atavistic images of something

32 David L. Pike, "Sewage Treatments: Vertical Space and Waste in Nineteenth-Century Paris and London," in Cohen, "Introduction," p. 51.
33 Pike, "Sewage Treatments," p. 55.
34 Cohen, "Introduction," p. xxii.
35 Cohen, "Introduction," p. xxii.
36 Pike, "Sewage Treatments," p. 60.
37 Pike, "Sewage Treatments," p. 52.
38 See Pike, "Sewage Treatments," p. 52.

"dark, womblike, warm and safe,"[39] quite the opposite of an ideal image of enlightened urban society. Renovation of the city's subterranean infrastructures is thus equated with an exorcising process of purification, cleansing, organizing the environment. In the descriptions of the Parisian sewers, the revolting underground is often exorcised through techno-utopianism – a procedure which can be seen as analogous to the Cartesian imagery of the body-machine, where the "natural body" becomes more graspable through the domesticating process of mechanization. The entrails of a dissected body translated into the language of mechanics, science and hydraulics are definitely less vast and more controllable. Likewise, the urban gut(ter) had to be exorcised and deprived of its horrific qualities. In scientific and popular narratives describing sewer workers and their working space such notions as filth, stench and disease are hardly ever mentioned. Instead, the focus is on the rationalizing discourse of idealized labour, utility and modern machinery,[40] thus in a way annihilating the vertical stratification of urban space with all its abject consequences. Another symptom of domesticating the subterranean abject was romanticizing the sewer and its workers, which could be found in a peculiar sewer mythology and with the sewer tour as an essential point of any Paris itinerary after 1867.[41] Pike writes: "[t]he Paris tour promised integration with the underworld in a system rational enough to be safe, but still organic enough to thrill."[42] The controlled thrill, processed through the discursive institution of paying for a ticket, performs the same domesticating function as contemporary horror films watched in hi-tech cinemas. The sewer tour's exorcising power lay in the fact that it was the triumphant spectacle of the technological curbing of the abyssal undergrounds. As a place of modern exploration, the sewers could no longer remain an unmapped no-man's land ruled by some half-man-half-beast figures excluded from the civilized world or a realm of some abject dream-like creatures like the Manhattan sewer alligators or monstrous giant rats.[43] It had to be appropriated, institutionalized and restrained, just like the sixteenth century dissected body. The procedures discussed here are just small examples of what seems a complex attitude underpinning modern utilitarian narratives. Mapping the body insides and domesticating the urban underworld, exploration of the underground sewers transformed into a spectacle and an anatomical dissection converted into a performance: they are all embodiments of the same process of appropriating disruptive qualities of the abject into the rational discourse of purposefulness.

---

39 Pike, "Sewage Treatments," p. 53.
40 See: Pike, "Sewage Treatments," p. 66.
41 Pike, "Sewage Treatments," p. 66.
42 Pike, "Sewage Treatments," p. 68.
43 See: Pike, "Sewage Treatments," pp. 52-53. The famous urban legends of albino alligators living in the sewers probably does not date back as far as the nineteenth century, as its first written accounts might be found around the 1920s, but – which is often the case with urban myths – such a possibility cannot be excluded.

Marcin Mazurek

Recycling the Visual: Hyperreal Practice and Rituals of Oblivion

> *"My God, don't they know? This stuff is simulacra of simulacra of simulacra. A diluted tincture of Ralph Lauren, who had himself diluted the glory days of Brooks Brothers, who themselves had stepped on the product of Jermyn Street and Savile Row, flavouring their ready-to-wear with liberal lashings of polo kit and regimental stripes. But Tommy surely is the null point, the black hole. There must be some Tommy Hilfiger event horizon, beyond which it is impossible to be more derivative, more removed from the source, more devoid of soul."*
>
> William Gibson, *Pattern Recognition*

The above monologue, performed by Cayce Pollard, the protagonist of William Gibson's 2003 book entitled *Pattern Recognition*, perhaps rather surprisingly for those accustomed to Gibson's dystopian urbanscapes, locates the reader in an even more ambiguous world – that of coolhunting, a professionalized search for the hottest trends and future fashions sponsored by multinational, anonymous and de-humanized corporations whose key managerial tasks require, in this case, a detailed semiotic analysis of a company's new logo. This is precisely a coolhunter's job – to design, redesign or simply evaluate the most intricate details of what is more than a mere recognizable trademark; the logo operates on the absolute front line of the corporate image war and, as Gibson convinces us once again, it is the image that decides about both the company's and the individual's present and future conditions alike.

In this uncontrollable, profit-oriented world of semiotic warfare in which the fittest are those whose identities are constructed more consistently than those of the others, the image becomes the critical component of both institutional and individual post-human identity, capable of either securing their social survival or sealing their fate. No wonder, then, that it takes a very gifted person, in this case coolhunter Cayce Pollard, to predict potential success or failure, the latter possibly resulting from the inappropriate curvature of the company's logo's initial letter. No surprise either that in Gibson's vision of the not-so-exaggerated present rather than a cyberpunk future, coolhunters become prophets of the new era who, due to their sixth marketing sense, indirectly decide the fate of millions. In other words, Gibson redefines, or perhaps simply exposes, the new topography of social existence in the post-capitalist environment, making an almost too obvious a claim that the ultimate goal and, at the same time, the most precious value is that of the complex and multi-layered concept of style.

Gibson's narrative, however, seems to be based on an implicit assumption that, contrary to the popular collocation, the style in question operates in an essentially self-reflexive manner, i.e. it is a style without substance, a pure style composed of detached signifiers released from the obligation to relate to the referential sphere through their conventionalized signifieds. The style in question is thus a dynamic intertextual phenomenon, drawing from, mixing and debating with other styles, languages and codes, regardless of history, chronology and locality as well as of possible value judgements, distinctions between high and low culture, or even the question of taste. Deprived of a clearly demarcated referential horizon, the style deconstructs its own singularity as its territory is not only open to related styles and fashions but in fact feeds on and is recognizable only through the dynamic network of these connections. And from this perpetual flux of textual and visual exchanges there seems to be no turning back, no singular context to be ever univocally located.

Such an approach inevitably evokes a number of critical post-structuralist and post-modern perspectives upon the nature of signification. From Roland Barthes' definition of the text as an interpretive field, to Jacques Derrida's exploration of textual peripheries, to Jean Baudrillard's postulated primacy of simulation, one might observe first a growing distrust in the reliability of the sign as such, and then a sort of steady theoretical agreement on the gradual separation between the sphere of signification and that of referentiality. Few claims seem to grasp this particular departure more vividly than that of Deleuze and Guattari, their postulated "principle of multiplicity."

Though originally used as a characteristic of a rhizome, next to principles of connection and heterogeneity, the principle of multiplicity operates as a crucial feature of the textual experience. And as such, even though initially applied to the notion of a book ("[A book] is a multiplicity"[1]), it appears just as applicable to that of the aforementioned notion of style understood as a self-referential signification entity, which, constructed of multiple sign systems

> ceases to have any relation to the One as subject or object, natural or spiritual reality, image and world. … A multiplicity has neither subject or object, only determinations, magnitudes, and dimensions that cannot increase in number without the multiplicity changing in nature.[2]

One of the passages from the book, very illustrative for the intertextual nature of fashion signs, depicts an ordinary office meeting which immediately turns into what may at first glance look like a typical female whose-got-a-better-dress

1 Gilles Deleuze and Felix Guattari, *A Thousand Plateaus*, in Julie Rivkin and Michael Ryan, eds., *Literary Theory: An Anthology* (Oxford: Blackwell, 2001), p. 514.
2 Deleuze and Guattari, *A Thousand Plateaus*, p. 517.

contest, and which is, in fact, a metaphorical example of the post-industrial image war:

> Dorotea's black dress, for all its apparent simplicity, is still trying to say several things at once, probably in at least three languages. Cayce has hung her Buzz Rickson's [jacket] over the back of her chair, and now she catches Dorotea looking at it.
>
> The Rickson's is a fanatical museum-grade replica of a U.S. MA-1 flying jacket, as purely functional and iconic a garment as the previous century produced. Dorotea's slow burn is being accelerated, Cayce suspects, by her perception that Cayce's MA-1 trumps any attempt at minimalism, the Rickson's having been created by Japanese obsessives driven by passions having nothing at all to do with anything remotely like fashion. ... It is an imitation more real somehow than that which it emulates.[3]

Such a shift of attention in determining identity of the Other, from social, bodily or facial features to their mere cover, inevitably brings about a number of questions revolving not merely around the reliability of identity determinants but first and foremost around the relation between the outer/visible and the inner/invisible. And even though the implied lack of a link between the outside and the inside clearly inscribes itself into the hermeneutics of universal suspicion so typical of post-modern culture, one more claim seems to be being made: in the context of heavily visualized modes of social existence, unassisted identity constituents, such as for instance the human face so much favored by Emanuel Levinas, lose both the battle and the war against the power of image constructed through a peculiar mixture of technology, pop culture and subtle intertextual connotations which have become legitimate post-industrial means to the ultimate end – the ability to endlessly construct and reconstruct one's own identity.

In Gibson's book identity reconstructions take place on several levels, the most obvious referred to as a "gender bait" – the practice of misleadingly introducing oneself via the Internet as a representative of the opposite sex (usually passing oneself off as an attractive female) in order to generate the desired reaction (from a naïve and unsuspecting male). Though hardly innovative, let alone prophetic, this gesture remains nevertheless a plain component of the simulation culture, implicitly pointing to a very close bond between fashion industry and the digital environment, the combination of which seems to have released the whole new array of identity re/construction possibilities, effectively turning the very process of identity formation into a repeatable ritual of identity selection.

Gibson seems thus to make one more implicit statement, once again bridging the gap between postmodern theory and literary practice. Flexible identity

3 William Gibson, *Pattern Recognition* (New York: Berkley Books, 2004), p. 11; brackets provided.

determinants, endlessly reconstructible through fashion traces, have long ago ceased to be singular signs identifiable through the relation with their signifieds. Today, Gibson seems to claim, within the post-alphabetic discourse of visualised identity components assisted by the ever-presence of visual technologies, there is no such thing as an isolated signifier but a complex network of textual-visual connections meant to be watched and absorbed rather than read and contemplated.

Arthur and Marilouise Kroker, somehow providing theoretical justification for Gibson's visions, see the reasons behind this situation in the rampant and uncontrollable development of media and electronic technologies, which redefine the postmodern condition and generate what they refer to as "the world of digital reality:"

> The real world of digital reality has always been post-alphabetic. Probably because the letters of the alphabet were too slow to keep up with the light-time and light-speed of electronics, ... The result has been the end of the Gutenberg Galaxy and the beginning of the Image Millennium. Images moving at the speed of light. Images moving faster than the time it takes to record their passing. Iconic images. Special-Effect Images. Images of life past, present and future as culture is fast-forwarded into the electronic nervous system. Images that circulate so quickly and shine with such intensity that they begin to alter the ratio of the human sensorium.[4]

Strangely enough for the world of perpetually circulating images, fashions and intertextual styles, what the book's protagonist is after is pretty much the opposite of what the general public is exposed to in every-day consumer practice. Taking the obsessive search for a new style, always-already composed of other styles, to an extreme, Gibson suddenly brings the techno-fashion machinery to an unexpected halt. This is because what Cayce Pollard is desperate to find is precisely a mysterious and de-contextualized signifier, pure in the sense of not negotiating anything, unidentifiable in terms of connotations, allusions and visual/textual exchanges with other complex signifiers. The signifier in question takes the form of a collection of enigmatic video clips anonymously uploaded to the Internet which begin to generate a growing net-wide cult, which is, typically for the era of global circulation, unattached to any particular location. What is unique about the clips is that even though they always depict a couple there are no indications whatsoever when and where the footage was shot and thus no point of reference, no context and no connection. As Cayce notices while watching the fragment no.135:

> He might be a sailor, stepping onto a submarine in 1914, or a jazz musician entering a club in 1957. There is a lack of evidence, an absence of stylistic cues, that Cayce understands to be utterly masterful.

---

4 Arthur and Marilouise Kroker, "Eye-Through Images," http://www.ctheory.net/articles.aspx?id=224 (Last accessed 15 Dec 2007).

> ... The girl wears a ... coat, ... its shoulder-padding the subject of hundreds of posts. The architecture of padding in a woman's coat should yield possible periods, particular decades, but there has been no agreement, only controversy. ...
>
> The one hundred and thirty-four previously discovered fragments, having been endlessly collated, broken down, reassembled, by whole armies of the most fanatical investigators, have yielded no period and no particular narrative direction. ...
>
> [And] Cayce knows that she knows nothing, but she wants nothing more than to see the film of which this must be a part. [5]

Much as within the contemporary Western experience defined by image bombardment and info-blitz, one might identify with Cayce Pollard's desperate longing for a space of pure originality, a space, "beyond which it is impossible to be more derivative, more removed from the source," her longing nevertheless seems thoroughly futile.

This is because, leaving the book for a moment now, in contemporary theory originality has been repeatedly pronounced dead or has at best been treated in terms of a sentimental myth, ironically enough exceeding even Gibson's prophesies. Whether we take Terry Eagleton's commonsensical remark that "[i]t is difficult to think of an origin without wanting to back beyond it,"[6] or a more elaborate approach, presented for instance by Walter Benjamin describing the disappearance of an artefact's aura among its countless reproductions, it seems that a new paradigm is now at work in the semi-virtual environment of the Western world, namely, the ecological urge not to let anything go to waste or be forgotten. In other words, to make sure that everything is reused or recycled. Jean Baudrillard summarises this particular cultural condition in the following way:

> The ecological imperative is that all wastes must be recycled. ... There is in fact no insoluble waste problem. The problem is resolved by the post-modern invention of recycling and the incinerator. ... We have to come to terms with the idea that everything that was not degradable or exterminable is today recyclable, and that there is no final solution.[7]

Interestingly enough, for Baudrillard the obsession and desire to repeat and re-enact is not, as one might predict, a symptom of the end of culture understood in terms of linear progress and controlled development. The immediate assumption might suggest that, since all that Western populations participate in, is a collection of recycled shows, ideologies and representations, culture has come to a halt. It is

5 Gibson, *Pattern Recognition*, pp. 23-24.
6 Terry Eagleton, *Literary Theory. An Introduction* (Oxford: Blackwell, 1998), p. 13.
7 Jean Baudrillard, "The Event Strike," in Jean Baudrillard, *The Illusion of the End*, trans. Chris Turner (Stanford, Ca.: Stanford University Press, 1994), pp. 26-27.

no longer moving forward but is precisely making a step back into its own past, resurrecting its bygone entities, waking the dead in a desperate attempt to appear fresh and alive while in fact generating zombie products which though they appear alive are technically dead or at least long gone. Such an approach, suggesting a state of cultural exhaustion and justifying the need to look backward rather than forward, could easily explain the popularity and ever-presence of remakes, prequels, remixes and sequels, from Puff Daddy's version of Led Zeppelin's "Kashmir" which, perhaps not surprisingly, served as a soundtrack for a remake of *Godzilla*, a Japanese collection of monster movies from more than twenty years ago, to, for instance, *Ocean's Twelve*, a mere sequel to *Ocean's Eleven*, itself a 2001 remake of the original 1960 movie.

For Baudrillard however, this cultural tendency, or perhaps even an industry, to resurrect the dead, is a sign of, if not progress, then at least a new stage in cultural evolution. The new stage is essentially archaeological, as its main concern is to revitalize all the fossilized artefacts and events, to make these fossils dance again. "All the relics," writes Baudrillard, "all the traces which were shrouded in the greatest secrecy and which, by that token, formed part of our symbolic capital, will be exhumed and resuscitated, ... we shall turn them from something buried and living into something visible and dead; we shall turn their symbolic capital into a folkloric, museum capital."[8]

Typically for Baudrillard the one to blame is technology, as it is precisely the potential of technology to first explore or excavate and then immediately visualise even the invisible through the invention of simulacra, and hence magically turn collective forgetfulness into collective memory. "The extraction of relics," claims Baudrillard, "has become an industrial undertaking. [...] Palaeontology is advancing at the same pace as the latest technology. Sites, relics, tools and bones: a whole stratum of signs, many thousands of years old, wrested back from oblivion. Thus technology can pride itself on enriching the cultural heritage ..."[9]

The only problem with such artificially recreated heritage appears to be one of quantity. In the world of autonomous images and replaceable visions it is the abundance and surplus of identity patterns that puzzles and bothers their viewer or participant. Baudrillard's notion of "enriched cultural heritage" seems much too mild a term. According to Zygmunt Bauman, such overproduction of identity patterns not only turns active subjects into passive consumers, but is a source of a new kind of existential pain for those who are meant to choose the most suitable identity pattern for themselves.[10] Paraphrasing William Gibson's title once again, to recognize a seemingly unique and apparently inimitable pattern is not an easy task in a culture in which the production and re-production of sign systems is an

8 Jean Baudrillard, "The Dance of the Fossils," in Baudrillard, *The Illusion of the End*, p. 72.
9 Baudrillard, "The Dance of the Fossils," p. 73.
10 Zygmunt Bauman, *Płynna nowoczesność* (Kraków: Wydawnictwo Literackie, 2000), p. 97.

"industrial undertaking." Acquisition or recognition of one particular identity pattern, even if it brings the desperate search to a momentary halt, is never bound to solve the problem, since, as Bauman is careful to note, the act of becoming somebody, which is bound to end the identity search, brings about the bitter realisation that being somebody is not quite equal to being oneself.[11]

Still, somehow contrary to Baudrillard, who sees the post-modern semiotic confusion as a result of technology-produced simulacra, Bauman sees the situation in the larger context of much more complex cultural operations. For Bauman, the Western world has entered a new phase of Modernity, rather than Post-Modernity, which is characterized by the multi-layered metaphor of liquidity. "Liquids," writes Bauman, "do not maintain their shape for a long time, but are always ready and keen to change it."[12]

Bauman locates the always-evolving liquid bodies in relation to solid bodies, pointing out the former's ability to dissolve the latter. These two conditions are used as powerful metaphors for the shift between early modernity, characterized by solid social structures, and the contemporary stage, which Bauman defines precisely as "liquid." Still, rather than replacing the solids, contemporary liquids seem to take advantage of their dissolving capacity and simply absorb the solids into the unlimited and reproducible melting space of contemporary culture.

One might naturally pause here to ponder upon the nature of such a liquid-permeated space. Much as for Bauman, the metaphor of liquidity signifies instability and the (often disinterested) co-existence of social and philosophical paradigms capable of "pouring" themselves out of or into the most unexpected places and contexts, it seems difficult not to draw certain parallels between Bauman's recognition of culture which is by definition fluctuating in barely predictable directions and Baudrillard's identification of the hyperreal, media- and technology-determined space of recycling and resurrection in which, to use his own expression, all the fossils dance again. Their dance however, appears not so much as the dance of relicts, but as that of duplicated sign systems filtered through the simulation industry capable of immortalizing them, though at the cost of their extra-visual significance. Relicts, Baudrillard seems to be claiming, long ago deprived of their meaningful aura, have turned into an endless collection of recyclable *motifs* involuntarily engaged in the game of their own simulation, narrating, like Cayce Pollard's jacket, a new understanding of a legitimate sign as "an imitation more real somehow than that which it emulates."[13]

Baudrillard seems in no doubt about the location of such a twisted cultural condition in the context of post-humanism. Provocatively enough, especially in the context of numerous "posts-" and "ends" which infiltrate a large number of

11 Bauman, *Płynna nowoczesność*, p. 97.
12 Bauman, *Płynna nowoczesność*, p. 6; my translation.
13 Gibson, *Pattern Recognition*, p. 11.

contemporary discourses, he does not announce yet another end – this time, as one might predict, that of signification – but a precisely opposite condition: that of the impossibility of narrating any final closure, whether that of culture or of discourse. Such an end seems to Baudrillard impossible for at least one vital reason. To announce it, one would have to allow for the end of the reproductive power of Western culture itself, the power and urge to reproduce its visual and textual artefacts *ad infinitum* through an ever-expanding galaxy of sign systems. Such an assumption is unacceptable to Baudrillard – partly because the good news of culture immortalizing itself is, upon closer look, dimmed by the sinister shadow of this very culture's inability to depart from its essentially recyclable nature. And nature, through an ongoing exchange between memory and oblivion, seems to have imprisoned itself in an vicious circle of repetition. Immortality and eternal damnation appear thus to go hand in hand in Baudrillard's oeuvre, paradoxically redefining the post-human condition precisely in terms of the impossibility of its attainment or realization. In other words, the End is never bound to take place because it had already happened, triggered by "the post-modern invention of recycling and the incinerator," which, through their ever-presence and mass orientation, have replaced the sphere of objects by their mediated simulacra, both recycled and recyclable in a process to which "there is no final solution." Baudrillard finds no other way to conclude his commentary on this condition, than through an infernal metaphor:

> This is what heaven – or hell – must be like: the massive recall, at every moment, of all the patterns of our life. The penitentiary immortality, the carcereal immortality of an unrelenting memory.[14]

14 Baudrillard, "The Dance of the Fossils," p. 73.

Marek Kulisz

Recycling and Culture

Combining the words "recycling" and "culture" will yield either a comprehensible phrase or a paradoxical one depending on our understanding of the word culture. Recycling in its common usage has a fairly clear and by now well-established meaning: it is the re-use or recovery of waste or scrap materials. But "[c]ulture is one of the two or three most complicated words in the English language. This is so partly because of its intricate historical development, in several European languages, but mainly because it has now come to be used for important concepts in several distinct intellectual disciplines and in several distinct and incompatible systems of thought."[1]

If we link the idea of re-use with culture in its broadest sense, i.e. the sense in which it is used by anthropologists as the opposite of nature, then joining culture with recycling will of course be logical. Recycling, though modeled on natural phenomena, is not a natural process. It is a series of carefully planned activities involving state-of-the-art technology, activities which have their origin in man's realization that nature cannot "cope" with a number of man-made materials. In such a context the phrase "culture of recycling," though rather vague, is acceptable. It may, for example, be understood as denoting a situation now typical of modern society, which produces so much waste that it is forced not only to develop new technologies to deal with it but also change people's attitudes, i.e. modify both their thinking and behaviour.

The reversed phrase – recycling of culture – is also acceptable but only metaphorically. We can, of course, talk straightforwardly of recycling the material objects of culture but in such a case recycling of culture would in effect denote the same activities as the culture of recycling, and obviously we cannot narrow down our analysis to those objects. We have to take into consideration the vast non-material sphere of culture which includes religion, art, philosophy, politics, science, etc. If, however, we would like to talk about recycling with reference to these non-material products of culture, we will first have to resolve three questions: firstly, we must decide if anything in the non-material sphere can be thought of as waste; secondly, we should consider if this waste needs to be recycled, as it may simply "decompose" in our minds; thirdly, we have to find out if this type of waste can be recycled – in other words, what recycling would mean in this context.

It seems possible to give positive answers to the three questions though, as has already been stated, metaphorically, to indicate certain resemblance. As far as

1 Raymond Williams, "Culture" in Raymond Williams, Keywords. A Vocabulary of Culture and Society (London: Fontana Press, 1988), p. 87.

the first question is concerned, we can say that there exists such a thing as intellectual waste or even intellectual garbage. Theories, plans, ideas, ideologies, etc. that proved to be mistaken can be regarded as such waste. For instance, in the realms of social and political theories, the doctrines of the totalitarian state – fascism and communism – may serve as good examples. And if we look for further similarities with material waste, we will also realize that the doctrines have to be recycled. First, because they are more "toxic" than the worst type of nuclear waste – communism and fascism led to the deaths of millions of people around the world and continue to do so. Second, they do not simply rot away – they did to some extent but still there are neo-fascist and neo-communist organizations even in democratic countries. The third question that remains to be addressed is the one about actions or activities that we can refer to as the recycling of intellectual waste, in this particular case as the recycling of doctrines and ideologies.

In order to find some sort of resemblance here we must remind ourselves of the purposes that recycling serves. There are basically two such purposes: one is to minimize pollution by getting rid in a safe way of toxic materials and those that do not decompose, or that do but not fast enough, e.g. plastics. The other purpose is resource conservation through the re-use of certain materials, e.g. scrap metals and returnable containers, but also through the use of easily decomposable materials that once seemed to be useless, as for example hay, which is burned as fuel to produce heat.

If, as has been written above, we can talk of people's minds as being "poisoned" or "polluted" by false and harmful ideas, then what minimizes such "intellectual pollution" can perhaps be called recycling. In this sense the toxic waste of totalitarian doctrines can be said to be recycled by/through fiction, documentaries, and works of art whose aim is to expose the evil inherent in these doctrines. Books such as *Nineteen Eighty-Four*, *The Gulag Archipelago*, *The Captive Mind*, *The Open Society and its Enemies,* and films such as *The Conformist*, *Schindler's List*, etc. could be thought of as attempts at recycling, as they prevent the doctrines from being disseminated (from penetrating into the "soil" of our minds) and in this way minimize the danger totalitarianism poses to society.

The other major purpose of recycling, apart from pollution control, is resource conservation. The question now to be resolved is whether it makes any sense to talk about resource conservation with reference to the non-material – mental, intellectual, spiritual – sphere of culture. Psychotherapists have long been telling us that psychic energy is limited in the same way as physical energy, and that we should (this is not a sarcastic remark) give our brains a rest every now and again. But this is common knowledge now and there is no point in analyzing it here. The interesting problem is whether resource conservation makes any sense with reference to the non-material products of the mind – ideas, beliefs, theories, etc. which have gone out of use. We can certainly find numerous examples of re-use in this respect but whether such re-uses could be called recycling in the sense of resource conservation is a moot point.[2]

In religion, for instance, the now widespread rejection and abandonment of Christianity have led in the past few decades to the revival of pre-Christian beliefs and rituals, and thanks to media coverage such phenomena have become rather conspicuous. Ancient, mysterious, abandoned places of worship such as Stonehenge have ceased to be just tourist attractions; now they are crowded with believers (though the word "believer" may not be an appropriate one in this context as it is unclear what or who those believers believe in, especially that frequently such revived rituals are no more than fads). The revival, i.e. re-use of abandoned forms of worship is a fact but in order to think of them in terms of resource conservation we would have to liken them to old, used containers for our spirituality which are still good enough to be re-used, and that it would be wasteful to discard. Such an opinion, though possible, is subjective.[3] It would obviously be unacceptable for an atheist, but also for a Christian, for whom Christ is the best and only acceptable "container." If a Christian accepted the metaphor of containers for spirituality, the abandoned, pre-Christian forms of worship would be like no more than shattered museum exhibits kept in glass cases – any re-use being out of the question. Depending on the point of view, then, the container metaphors would belong to two different realms: to recycling on the one hand, to archeology on the other.

Interesting examples of the re-use of old, discarded ideas can be found in literature. The name that immediately comes to mind is Jorge Luis Borges, whose frequent references to long-forgotten ideas and theories became a distinguishing characteristic of both his fiction and his non-fiction. Let us just mention one of his short stories, "Averroes's Search," in which Borges writes about the mediaeval Arab scholar's abortive attempts to analyze parts of Aristotle's *Poetics* which are devoted to drama. Since drama was unknown in Islam, and there was nothing in Averroes's knowledge to give him a hint and guide him towards understanding the idea of the theatre, he is doomed to fail. His efforts are wasted, his conclusions utterly wrong. It is easy to see the story as a metaphor of recycling, especially since Borges declares openly: "In the foregoing story I tried to narrate the process of a defeat."[4] The word "defeat" refers, among other things, to the fact that this

---

2 At this point we should, perhaps, mention what Victor Shklovsky – in his famous essay "Art as Technique" – calls "the law of the economy of creative effort," and which he criticizes fiercely. Apart from a few Russian advocates of this theory, Shklovsky also quotes from Herbert Spencer and Richard Avenarius. I do not include the theory in my essay as the economizing it speaks about concerns the process of reception, i.e. the mental effort of the reader/hearer – it does not speak of any conservation of the material of art.

3 The problem of subjectivity of perspective in the context of recycling is a very important one and it requires some clarification. This will be done towards the end of this essay.

4 J. L. Borges, "Averroes's Search," trans. J. E. Irby, in J. L. Borges, Labyrinths (Harmondsworth: Penguin Books, 1976), p. 187.

particular aspect of Averroes's research was worthless in the sense that it did not constitute any advancement of knowledge, which Averroes had expected it to do. Borges takes this "intellectual waste" to "fuel" his own narrative, through which he shows that there is a different use to Averroes's erroneous conclusions, as they are an end result of an effort which, though useless on the logical/cognitive plane, is of high ethical value: "Few things more beautiful and more pathetic are recorded in history than this Arab physician's dedication to the thoughts of a man separated from him by fourteen centuries."[5]

One can give numerous similar examples but they would not add much weight to the argument, as the only purpose here is to show that it does make sense to talk about recycling of culture, and the examples already provided prove this sufficiently. So far our analysis of recycling in culture has concerned the concept of culture in its broadest meaning, in the context of which both phrases – culture of recycling and recycling of culture – seem to be logical. But they will not be so when we take the word culture in its narrow but at the same time "[t]he most widespread use: [in which] culture is music, literature, painting and sculpture, theatre and film."[6] It is easy to see that in this sense culture is almost synonymous with art, and certainly there is much overlapping between the two concepts. Combining this meaning of culture with recycling will be illogical, if we use the term recycling in the meaning given to it at the beginning of this essay. The explanation why the two words cannot be combined in a logical phrase is quite simple.

Recycling – let us repeat the definition – is the re-use and/or recovery of waste/scrap materials. The meaning of the term is not just a result of the simple grammatical operation of adding a prefix and a suffix to a verb: re + cycle + ing. It is not any *re*-use, any *re*covery, any *re*-circulation, etc. And we should not be misled by etymological affinities, either. In order to talk about recycling we must first identify the waste material to be recycled. But identifying waste may not be possible when we think of culture in the narrow sense: What waste could we possibly find in the works of Michelangelo, Bach, Mozart, Rembrandt, Velázquez, Dostoyevsky, Kafka, and all the other great masters who have shaped our culture?

By asking the question we have for the second time touched upon the problem of subjectivity of perspective and perception, and we must briefly deal with it before we go on. Any individual may give a positive answer to the above question. Especially, a lover of the avant-garde could argue that there is at least some used-up, outdated material, if not waste, in the works of the old masters, in the sense that what they created no longer appeals to modern man's taste and sensibilities (A good example of such an opinion is Virginia Woolf's blistering attack on the Victorian writers in "Mr. Bennett and Mrs. Brown"). But such a view

5 Borges, "Averroes's Search," p. 181

6 Williams, "Culture," p. 90. "A Ministry of Culture refers to these specific activities, sometimes with the addition of philosophy, scholarship, history."

is of relatively little importance in the context of recycling, for at least two reasons: firstly, it is not uncommon that individuals change their opinions during the course of their lives, and their tastes change as well; secondly, and more importantly, a subjective perception is too narrow, as recycling denotes a series of activities undertaken on a much wider plane: by society or in the name of society. The opinions and actions of the individual become significant only in the wider, social context, and it is highly improbable that any society as a whole would agree on dismissing any of its great masters of the past.[7] In order to explain the notion of wastelessness in culture (in the narrow sense of the word, in which it is close to art), let us refer to T. S. Eliot's well-known essay "Tradition and the Individual Talent." Though the word culture does not appear in his essay, it is obvious that by tradition Eliot means cultural tradition, understood as a collection of works by the great, acknowledged masters. Eliot was probably the first writer who made us realize clearly that the idea of progress is not applicable to art because "[*a*]*rt never improves*, [only] the material of art is never quite the same."[8] According to Eliot we should be aware that "[t]he mind of Europe [...] is a mind which changes, and that this change is a development which *abandons nothing en route*, which *does not superannuate* either Shakespeare, or Homer, or the rock drawing of the Magdalenian draughtsmen. That this development, refinement perhaps, complication certainly, is not, from the point of view of the artist, any improvement."[9] And earlier in the same essay: "The existing monuments [of cultural tradition] *form an ideal order* among themselves."[10]

If we accept Eliot's views and agree with him that cultural tradition is some sort of mechanism or dynamic structure which does not improve, which does not abandon anything, and which forms an ideal order, then we must also accept the straightforward conclusion which follows from his propositions: the system does not produce any waste. If so, then talking about recycling in this context does not make sense. But nowadays there are lots of people to whom Eliot is no longer an authority. Especially to partisans of (post-)modern theory Eliot's notions of literature, art and tradition will be outdated and of little significance in our post-modern world. But, interestingly enough, modern theory, which has been founded predominantly on semiotics, only confirms Eliot's views in this particular respect. It tells us that culture (no matter how we understand the word) is a signifying system in which everything signifies. The system is free of any redundancy whatsoever, let alone waste. In literary studies probably the best-known analysis so

7 Notable exceptions are highly ideologized societies. But in a historical perspective such societies are short-lived: even the Nazis with their burning of books and the Communists with the truly Orwellian apparatus of censorship did not achieve their goals.

8 T. S. Eliot, "Tradition and the Individual Talent," in Selected Prose of T.S.Eliot, ed. Frank Kermode (London, Boston: Faber and Faber, 1980), p. 39; italics mine.

9 Eliot, "Tradition and the Individual Talent," p. 39; italics mine.

10 Eliot, "Tradition and the Individual Talent," p. 38; italics mine.

far that is a practical application of this theory of complete signification is Roland Barthes's *S/Z*, in which Balzac's short story is taken to pieces and analyzed sentence by sentence: nothing is omitted, there are no "leftovers," no "waste," nothing to be recycled.[11]

We can even argue that in its totalizing approach modern semiotics not only does not refute Eliot's theory of cultural tradition as an ideal order, but actually makes the theory radical in the sense that what Eliot wrote about abandoning nothing in culture in the narrow sense, semiotics applies to culture in the broad sense. As a result of this totalizing approach all other arguments and perspectives become invalid. Waste is a relative notion, which means that when we discussed "intellectual waste," we did it from certain perspectives. The waste of totalitarian doctrines was, for example, considered as such from ethical, economic, and logical perspectives: the doctrines proved to be wrong in at least these three aspects. Waste in scholarship was analyzed from the logical and cognitive perspectives: Averroes in his study of Aristotle's *Poetics* was simply wrong in the same way in which, for example, Ptolemy was wrong when he constructed his model of the universe. But in semiotics such considerations are beside the point because everything signifies – full stop. We see clearly now what we could predict at the beginning: that in order to talk in any sensible way about recycling in culture we must first try to explain what we mean by culture. The term is so complex that if we do not attempt to clarify its various meanings we may end up not knowing what we are talking about, especially that in some cases by linking culture with recycling we can even coin phrases that are self-contradictory. But in the instances in which recycling in culture appears plausible, new areas open up for analysis, which makes us realize that if it is possible to talk about mental hygiene, as psychotherapists do, it may also be possible to talk about ecology of the mind.

11 The practice of such detailed, "complete," analyses is of course by no means new. The Bible, for example, has for centuries been studied this way. In the Rabbinic tradition even single letters are endowed with meaning.

Carl Humphries

Metaphysics, Critical Theory, and the Illusion of Cultural Self-Reproduction

The idea that we live in a culture that reproduces its distinctive values, ideas and patterns of behaviour in ways not fundamentally accountable to human beings themselves is central to much theorising about contemporary society. Such theorising may set out to describe the social and cultural conditions of "postmodernity," or may reflect a self-reflexive theoretical stance that is distinctively "postmodern," in that it begins not from a description of how things apparently are, but from an acknowledgement of the implications for theory itself of the idea that all culture, including theorising itself, is self-reproducing.

These two starting points, though distinct, give rise to approaches that often merge into a broader view of all human culture as essentially self-reproducing – one that nevertheless takes our current situation as paradigmatic, in virtue of the destabilising effects of the information revolution and globalisation. Given the speed with which elements of our culture can be disseminated around the globe, it is hard to think of them as standing in stable relationships to the situations in which we actually meet with them. In that case it can be tempting to think that cultural phenomena do not derive their significance from our encounters with them – from their place in the pattern of our lives – so much as from their place in some more arbitrarily instantiated abstract system. If that were so, then we might also think similar considerations apply to the processes that determine which cultural forms persist. In other words it would look as if culture – or at least our contemporary culture – is self-reproducing.

Yet the claim that *our* culture is like this, and the claim that all culture is like this, are still, in principle, distinguishable. What makes them hard to disentangle in practice is that the second claim often emerges from a kind of self-reflexive theorising about what it means to oneself be a part of the culture one is seeking to understand, a kind of theorising, modelled on "hermeneutics," that seemingly has the potential to encompass and disarm any counter-claim that might be made, to the effect that not all culture is necessarily self-reproducing. For example, it can be argued that such a counter-claim can never be taken seriously because we can never lay to rest the thought that maybe our inclination towards such a view of culture as non-self-reproducing might *itself* be just an instance of culture reproducing itself, in that it happens to correspond to the traditional understanding of culture that has held sway *within* our culture.

If the idea that not all culture need be self-reproducing is prevented from gaining a proper foothold as a result of moves like this, then it is also difficult to elaborate reasons why one might wish to admit that *our* culture *happens* to display self-reproducing traits, while denying that culture in general is always inherently self-reproducing. Yet we might have very good reasons for wishing to adopt just such a position. For instance, we might feel that there are aspects of what it is like to be in a self-reproducing culture such as ours that are undesirable, and which ought to be changed. Obviously, articulating such a project of social change would be futile if what we take to be problematic about our current cultural situation were a necessary feature of all cultures. The problem, then, is that articulating such critical thoughts about our society today would mean challenging the claim that all culture is self-reproducing, and this is difficult precisely because the latter has been elaborated self-reflexively from the "hermeneutic" perspective. This perspective appeals to the widely felt intuition that human beings cannot in any radical sense transcend the cultural framework they find themselves in.[1] But someone seeking to develop a critique of the self-reproducing character of our culture, by asserting that culture need not always be like this, is required to go against this intuition. They must admit that they themselves also inhabit *our* culture – which in this case they do take to be, in some sense, a culture that tends to reproduce itself – while nevertheless denying that their own position is merely an instance of this self-reproduction. It is hard to see how this denial can have real force unless a "culture-transcendent perspective" is implicitly or explicitly attributed to the person who makes it.[2]

One problem, then, is that invoking such a perspective goes against the intuition just mentioned. But this intuition by itself is rather vague, if only because it does not spell out any precise view of what culture is, or why we should think of it as placing limits on how we see things. The more serious problems come into view when this sort of intuition is used to motivate a broader critical stance towards society that sees the idea of "culture-transcendent perspectives" as part of a larger framework through which human societies are able to sustain undesirable social conditions under a guise of normality. This stance is familiar to many as a

1 In fact this line of thinking goes beyond what is implied by the original use of hermeneutics to characterise the horizon-bound character of human life, which first and foremost concerns our understanding of the existential possibilities afforded by the *world* we find ourselves in: its hermeneutic ontology (as in Heidegger). The assumption here seems to be that *all* differences of perspective – and therefore cultural differences too – function as horizons in this same way. This collapses the distinction between existing in a world and participating in a culture in a way that may seem natural to those seeking to distance themselves from traditional Western metaphysics. However, the original hermeneutic model does still leave room for such a distinction to be made – albeit in terms less straightforward than those offered by traditional metaphysical thinking, and yet it, too, is supposed to represent an alternative to traditional metaphysics. This suggests that the relations between metaphysical and hermeneutic theorising are more complex that cultural theorists tend to acknowledge.

2 The use of the phrase "culture-transcendent perspective" may be misleading, as for the person who embraces such a "perspective" it may not, in fact, be a "perspective" at all, but simply "how things are." I therefore use this term advisedly.

consequence of the influence of poststructuralism in areas such as literary theory and cultural studies. It sees the notion of a "culture-transcendent perspective" as part of a framework of assumptions perpetuated by Western metaphysical thinking, used by societies to generate a veil of neutral authority.[3] Theorists such as Derrida and Foucault sought to disrupt this framework, chiefly by questioning two ideas that seem to underlie metaphysical thinking itself. These are the idea that signs necessarily possess a stable, collectively shared significance, from which we may reliably infer the communicative intentions of a speaker or writer, and the idea that forms of human discourse can be used to articulate and transmit knowledge in neutral terms, constrained only by impartial norms of objectivity and rationality.[4]

Both of these critiques imply that *if* we ourselves *were* being deceived or manipulated through culture-transcendent claims made on behalf of human language and knowledge, we would not be in a position to know it. This makes it seem as though the only claims we can safely accept are ones that self-reflexively take account of this very possibility by excluding any culture-transcendent perspective in principle. Yet this closes down discussion about whether culture in

---

3 In this paper I employ terms such as "Western metaphysical thinking," "traditional metaphysics," etc. as loose labels that refer to nothing more than what cultural theorists have in mind when they seek to distance themselves from any kind of thinking concerned with establishing metaphysical foundations in the way that they hold to be characteristic of the Western philosophical tradition. How far these labels really capture any meaningful features of that tradition is an open question.

4 A rough summary of the relevant points in the work of these two theorists may be useful here. For Derrida, instead of being necessarily stable and shared, meanings are inherently unstable – and so not shared. This is a consequence of the idea that a speaker or writer who entertains thoughts and intentions in their conscious or unconscious mind, before expressing them publicly, nevertheless uses concepts belonging to a system articulated in terms of *signs*, whose meanings are assigned with reference to relationships between signifi*ers* as well as signifi*eds*. If signifiers convey thoughts or enact intentions in the public realm, they necessarily have an existence outside of the mind of the speaker or writer, so for any particular signifier it must be possible that its spatio-temporal conditions will change away from those in force when the original assignment of meaning was made in the mind of the speaker or writer. Such changes seem to bring with them the unavoidable possibility that our own encounter with a sign takes place in an interpretative context different from that in which the original assignment of meaning was made, and this thought suggests that there will always be potential ambiguities in how our encounters with signifiers relate to the formal structures that determine meaning at the level of the signified (see J. Derrida, "Structure, Sign, and Play in the Discourse of the Human Sciences," in J. Derrida, *Writing and Difference*, trans. Alan Bass (London: Routledge, 1967)). For Foucault, on the other hand, the guiding thought is that the discourses through which knowledge is articulated privilege certain perspectives and interests over others when they present certain ways of thinking as transparently valid or true. They do this because they require us to implicitly accept various norms of thought as unproblematic, where these in turn transmit hidden assumptions about the world. This means that discourses in effect turn the knowledge-claims articulated through them into self-fulfilling prophecies, as these hidden assumptions influence our perception of reality and lead us to act in ways that make the world correspond more closely than it otherwise would to how these knowledge claims portray the world as being (see M. Foucault, *The Order of Things: An Archaeology of the Human Sciences* (London: Routledge, 1970)).

general is, in fact, self-reproducing or not, and as I have already suggested, there may be reasons for wanting to articulate a position that regards our culture as self-reproducing, but not all cultures as being so. It looks, then, as though we ought to try to see whether an alternative theoretical basis can be found that would allow us to at least keep discussion in this area open and ongoing.

Perhaps a good place to start is to note the parallel between this scenario and the wider sense of frustration that has emerged amongst those who feel that cultural relativism, when taken too far, ceases to serve the purposes that originally made it seem attractive as the world-view of choice for liberal first-world intellectuals. The perception that political correctness, in seeking to guarantee a non-discriminatory arena for public affairs, actually closes off debate about issues relevant to a multicultural society, the appeal to the rhetoric of cultural relativism by Bush and Blair as a justification for a globally imposed cultural and economic liberalism, and the not wholly unsophisticated use of cultural relativist arguments by right-wing organisations such as the British National Party (not to mention its employment to provide a rationale for honour-killings and female circumcision in non-first-world societies) – all these things have certainly muddied the waters, unsettling the perception that secular liberal individualism and cultural relativism share a common front in opposing absolutist dogmas, whether these be products of rationalism or of fundamentalism. Yet, as one reviewer of a recent anthology devoted to dissenting voices within cultural theory pointed out, nobody has proposed an alternative position that overcomes these paradoxes, while also continuing to address the issues that motivated critically oriented cultural relativism in the first place.[5] Critics of contemporary cultural theory and cultural studies tend to just take its internal inconsistencies as de facto evidence of a need to resurrect either traditional common-sense metaphysical realism or some set of pragmatic ideals that appeal to the apparently self-validating character of rational norms and scientific practices. The problem is that because these positions do not address the concerns of critical theorising about culture on its own terms, they do nothing to disarm the suspicion that what they really are is a return to the old authoritarian belief systems, to be used once again as tools for social repression by whoever happens to find themselves holding the reigns of economic and political power. Here, too, we see a kind of standoff developing, between those willing to accord culture-transcendent status to at least some forms of social normativity, because they see this as a necessary basis for any positive action in the world, and those who insist that the self-reflexive logic of hermeneutic theorising entails a principled scepticism about all culture-transcendent claims.

One way to break this deadlock would be to show that the self-reflexive critical stance adopted by many cultural theorists is internally problematic in ways

5 The anthology reviewed in this article is *Theory's Empire. An Anthology of Dissent* (NY: Columbia University Press, 2006). See S. Jarvis, "Thinking about Theory," in *The Times Literary Supplement*, January 11, 2006.

similar to those which these same theorists claim are features of Western metaphysics – providing, of course, that one can do this without presupposing the validity of Western metaphysics itself.[6] For example, just as critical theory argues that Western metaphysics and its discourses generate the illusion of their own culture-transcendent validity, so one may argue that deconstructive and Foucaultian discourses generate the illusion that they are the only real critical response to the metaphysical tradition itself. The thought is that in seeking to disrupt metaphysically implicated discourses, these critical approaches are forced to become examples of the sort of artificially self-validating discourse they aim to criticise. But how could one show this without prejudging the issue of whether metaphysical discourses are, in fact, deceptive in virtue of claiming culture-transcendent status for themselves? And where would this then leave discussions about whether culture is inherently self-reproducing or not?

I want to draw on some ideas from Wittgenstein, firstly to show that this kind of internal undermining of the self-reflexive logic of critical theorising is possible, and secondly to suggest that it has implications for how we should approach the issue of whether culture is self-reproducing or not.

In his *Tractatus Logico-Philosophicus* Wittgenstein (1922) explores a model of what it means to think of the world as governed by logic – as it would be if the basic norms of human rationality that constrain much of our everyday use of language turned out to have metaphysical significance for us after all. But he is also known for having later rejected this model, in favour of an approach often characterised as a rejection of all foundationalist theories of truth and meaning.[7] I propose that we treat Wittgenstein's early work as a thought-experiment that will help us to understand what it means to embrace a metaphysical perspective on the world, for someone who actually *does* embrace such a perspective.[8] This can, I think, enable us to see why the kind of critique of metaphysics that has been pursued by cultural theorists under the mantle of poststructuralism can only approach its own critical goals at the cost of becoming another metaphysical discourse, just like those which, if its description of their social implications were correct, it would be right to criticize. Of course the fact that this makes such

6 For reasons that should become clearer in due course, I do not wish to imply any position here about whether Western metaphysics is, in fact, "valid" or not. The point is just that a critique of a position that itself rests on a rejection of Western metaphysics will not cut much ice with proponents of the latter if it is seen to itself rest on Western metaphysics.
7 Nevertheless, there are also important points of continuity between these two phases in his philosophical development, and he himself claimed that his later work could only be understood properly in the context of what he had earlier tried to achieve.
8 One can think of this as another instance of self-reflexive hermeneutic theorising, parallel to that employed by critical theorists, but instead of asking questions about what it would be like *for me* if the world *that I myself am in* were to be made to appear to be governed by logic without really being so, we ask questions about what it would be like *for me* if the world *that I myself am in* were to *in fact* be logical.

cultural theorising self-contradictory cannot count as a fatal objection for its adherents, since they treat all norms of rational consistency as suspect anyway. But linking this to the implications of Wittgenstein's later work will, I hope, point to ways in which such theorising itself produces self-validating circularities like those it ascribes to cultures that embrace metaphysics and rationality, with the consequent implication that it is itself complicit in blinding us to possible forms of manipulative social repression.

The essential point made in the *Tractatus* is that when we seek to grasp the idea that there are definite, yet absolute limits to our reality (or the thoughts we can have about it) corresponding to the logical constraints governing how we think and talk, we lapse into nonsensicality. This is because the very idea of such limits *being* limits – in a sense that would grant them some definite character – would only make sense from a perspective that had also transcended those limits and seen what it is that specifically lies beyond them. Were those limits to be absolute – as they would be if logic "cut the world," so to speak, "at its joints" (in the way that traditional metaphysical philosophising held to be the case) – then this is exactly what we would *not* be able to do. The thoughts that would be rendered unthinkable by logic if logic were valid would be ones that we could not then think, so we cannot ever know which ones – if any – they will be.

For Wittgenstein we are liable to be deceived here by a false picture embedded so deeply in the conceptual grammar of our language that we easily fail to notice it. We speak of logic and its manifestations in language as imposing limits on what we can think or say about the world, where these limits are, at the same time, taken to correspond to limits on how the world itself can be. Yet the idea that these limits are "imposed" itself suggests that there is something beyond them – as there *would* be if, indeed, we were speaking of limits in other, more ordinary kinds of context. And this implies the opposite, which is that logical constraints do not, after all, define the limits of reality, but only those parts of it accessible to us. So we are using the term "limit" in a way that indicates lack of clarity about how its meaning functions in different circumstances.

For Wittgenstein, this is responsible for the illusion that something meaningful is being achieved when philosophers try to spell out how and why certain thoughts and sentences are excluded by logic, as nonsense, while others are not. But the relevant point for us is that it also creates another illusion. It suggests that we can grasp how things are for someone who embraces certain metaphysical truths and upholds the validity of some form of logic, at the same time as ourselves just regarding their way of perceiving the world as one of many possible partial perspectives on reality. The illusion is that we can move freely between these two sets of circumstances and treat them as commensurable. We might think of what they take to be reality and what we take to be just a perspective on reality as corresponding to a single set of limits under different interpretations.[9] Our critical scepticism towards someone else's unconditional beliefs and commitments can then seem like a way of encompassing their more naïve point of view within our own, so that a sceptical, relativistic stance somehow seems to subsume or supersede the alternative. For Wittgenstein, though, this would be confused nonsense, since talk of metaphysics and logic as these are for the person who believes in them, and talk of a cultural perspective or world-view of which these are merely the expression, simply pass one another by. Connecting them up would imply a single conceptual framework, capable of accommodating the limits to reality specified by metaphysics and logic, both as defined *only* from within (as absolute limits with no "other side"), and as defined from *both* within *and* beyond. The point is that when we spell out this line of thinking in this kind of way – i.e. in terms that avoid the confusions concealed behind the grammatical ambiguities of ordinary language – all we are left with is nonsense.

We can now see that a cultural theorist who uses this self-reflexive, "hermeneutic" reasoning faces a dilemma. As we have seen, this reasoning is relevant to the issue of whether culture is self-reproducing or not, as it can be used to suggest that a relativistic cultural agenda is the only appropriate response to the thought that we may all be systematically deceived by our own culture's traditional embrace of metaphysics and logic. But this leads to a principled rejection of culture-transcendent perspectives, and that in turn problematises the thought that even if our culture is self-reproducing, not all culture need be. Yet the problem for the cultural theorist is that one cannot reasonably adopt this self-reflexive, hermeneutic approach for oneself and not grant the right to others to adopt it too: indeed the logic of the relativistic agenda of the cultural theorist itself demands that one do this. But if the person who unconditionally accepts certain metaphysical claims and rules of logic is permitted to reason self-reflexively, they will come to precisely the conclusions registered by Wittgenstein in his *Tractatus*, from which, as we have already observed, no basis can be derived for conceiving of metaphysical frameworks and critical perspectives as at all commensurable.

This leaves cultural theory's attempt to dislodge or disrupt traditional metaphysics as part of a wider critical stance towards modern society looking very much like a toothless monster. It removes any basis for thinking that one could persuade people that certain problematic aspects of our society or culture are not

9 One might say: "under competing meta-perspectives." However, this would perpetuate the confusion, as it suggests that what the sceptic rejects is an equivalent but opposed meta-perspective to his/her own – one that consists of a second-order unconditional acceptance of one's unconditional acceptance of metaphysics and logic. This talk of *meta*-perspectives would itself imply that one's first-order unconditional acceptance is itself just a perspective. But that is to *already* describe it from the standpoint of the meta-perspective of the sceptic, and is certainly not how it looks from the other person's standpoint. Hence a false impression of commensurability is once again generated by this way of using language. Even talk of "competition" is misleading, as it implies a kind of commensurability.

part of the eternal scheme of things and so can and should be changed, as that would require one to insist on the superiority of a critical attitude towards traditional norms of thought over an uncritical one, which would imply commensurability between these two kinds of attitude. Yet the only alternative is to insist on this commensurability as a matter of principle, for the sake of the critique that it then seems to make plausible. The problem with this is that as we have seen this insistence implies that one is willing to refuse others the right to also make use of this kind of self-reflexive hermeneutic reasoning, as and where it leads to conclusions other than those one would personally wish for, suggesting that the latter conclusions themselves amount to unconditional truths. Hence the price for still thinking there can be a critical engagement between cultural theory and traditional metaphysics is acceptance of the transformation of the self-reflexive, hermeneutically elaborated version of cultural relativism into another absolute metaphysical framework.

One can try to differentiate this from traditional metaphysical thinking by asserting that the decision to embrace certain unconditional certainties is motivated here by *strategic* considerations that will also mitigate its potential for causing the social problems in which traditional metaphysics discourses have been implicated. These strategic considerations might be seen as directly implied by certain ethical intuitions or ideals that no reasonable or civilised person would question.[10] Yet this latter thought has no value, once we realise that the *strategic* implications of such intuitions and ideals can only be *actually* determined by asking how we should think or behave, in the light of these, against the background of some more or less determinate general understanding of the world, of human nature, etc.[11]

Another important consideration that speaks against this position is that even if it could be shown that it did not perpetuate or normalise *those* social problems, it is closely associated with a supporting framework of ideas about the nature of language and the place of human beings in the world that, if embraced in conjunction with this, will do the same, albeit for *other* problems. Insisting that a critical attitude towards metaphysics and culture is in principle superior to an uncritical one seems a lot less ad hoc when accompanied by the kind of poststructuralist views about the nature of meaning and knowledge mentioned earlier. These views lend support to this, but only because their own plausibility as general accounts of how meaning and knowledge are actually constituted depends on a set of implicit Cartesian dualist premises, and in fact it is these premises that make it more appealing to think of individuals as being in need of emancipation from the constraining effects of particular world-views than to think of them as depending on the latter in some more absolute sense for the basic form of their

10 This seems to me to be the basic idea of what has come to be known amongst cultural theorists as "strategic essentialism."

11 Ethical ideals and intuitions in a vacuum have no strategic implications, partly because there is nobody for whom they could have them.

existence. One reason why it is hard for cultural theorists to grasp this is that it requires us to first become aware of the manifold ways in which our own involvement with language tends to proceed with such premises being in place, and there is no way to articulate this thought within the poststructuralist framework itself, precisely *because* it implicitly takes these premises for granted. So we need an alternative approach to language that rigorously refuses to allow such premises to take hold from the very start. This is where I think Wittgenstein's later work has something useful to offer.

Wittgenstein's investigation of the nature of our language and concepts is guided by the intuition that for these to make any difference to our lives, they must be acquired in ways that are not simply arbitrary. This non-arbitrariness implies the possibility of coming to recognise, for something that we might want to consider as being a sign, the difference between meaningful acts of sign use on the one hand, and arbitrary or accidental appearances of the same underlying phenomena on the other. Put simply, this – for Wittgenstein – requires us to recognise that the former, but not the latter, stand in a special relation of "fit" to the context in which they are encountered, where this relation of "fit" is too specific to be explained by anything other than the thought that the sign is being used intentionally, to *mean* something.[12] Hence the question of what it is for something – or anything – to have a meaning at all, is only really answerable relative to these contexts. Since these contexts are utterly diverse, and the relation of fit that the use of a sign can stand in relative to these can vary enormously, no standard account can be given of what it means for something to have a meaning. If that is right, then there is no standpoint from which one could substantiate the central postulate of structuralist and poststructuralist accounts, which is that meanings are constituted from the outset relative to their place in an overall system of differences of meaning.[13]

The crucial implication of this is that to arrive at the sort of conceptual frameworks needed to articulate an understanding of reality – and of our place in that reality as individuals – we must first recognise such relations of "fit," but recognising these relations means being *already* committed to a determinate understanding of the relevant aspects of the surrounding context that, along with the instances of sign use themselves, constitute the elements of such a relation. In other words, contrary to what both structuralism and poststructuralism have suggested, a determinate relationship to the world is required to be in place in

12 This seems to me to be the motivating thought behind his insistence that for something to be identified as an instance of sign use, and thus as involving things which we might wish to call signs, it must stand in a reidentifiable relation to a set of public criteria. This highlights the fact that the things we call signs do not themselves normally furnish the basis for these relations, but must be embedded in regularities of human behaviour in order to do so – regularities that, for Wittgenstein, frequently (but *not* always) correspond to our *use* of the signs for particular purposes.

13 See especially the first part of Wittgenstein's *Philosophical Investigations* (Oxford: Blackwell, 1953).

advance of our discovery of the possibilities afforded by linguistically expressible concepts. At the same time, because this relationship is determined presuppositionally, as a precondition for acquiring a grasp of particular forms of meaning, it is defined relative to those particular forms. So if the ways in which signs become graspable as meaningful – and thus *as* signs – are utterly diverse, we can only state what is presupposed by particular forms of linguistic understanding, and no absolute model of reality or human understanding is implied. The Wittgensteinian approach is thus at least as resistant to dogmatic assertions of a single overarching metaphysical system characterising reality as poststructuralism, but avoids the implication of the latter that *all* normative perspectives are relative to a formal system of meanings whose relationship to anything beyond itself – such as, for example, the circumstances in which individual cases of sign use occur – must, as a result, be lacking in any form of normative predictability. A Wittgensteinian analysis can tell us what elements of reality are presupposed as having been taken at face value by us, *if and when* we engage in certain linguistic practices, but it does not oblige us to actually partake in any such practices in an absolute and unconditional way, and so is immune to accusations of being potentially hegemonic. At the same time, though, it leaves no room for the idea – central to poststructuralism – that language or culture, as an overarching system of meanings or concepts, could come *between* the individual and the world and mediate the entirety of this relationship.

This view of language and meaning can throw into relief the way in which Derridean and Foucaultian critiques depend on Cartesian presuppositions, as well as suggesting reasons why, given that this is so, we should be wary of embracing the kind of self-reflexive relativism about culture that, as we have seen, figures in debates about the self-reproducing character of culture. In Derrida's case, the destabilisation of meaning that is thought to occur when signs are recontextualised is taken as a necessary consequence of their intrinsic iterability. Yet this only follows as a general consequence if we assume, with Derrida, that stability of meaning, were it to exist, would have to consist in a knowable correspondence between what a sign comes to mean in various public contexts and what it once meant for its user, relative to the formal structures of meaning that determine a particular private intentional context. Put this way, meaning can only ever be unstable, but this is only against the background of a particular model of what it means for meanings to be *stable*. Given the fact that we spend a large part of our lives acting as if meanings were stable, the basis for invoking this model cannot lie in the thought that it makes more sense of our actual behaviour. Instead, it must reflect some deeper intuition that what ultimately matters about our relationship to the signs we use is how we stand towards them as individual subjects – how those signs figure in our conscious and unconscious mental processes – even when this has no bearing on how our use of them will be read by others. The point is that this intuition is *only* credible if we *already* have a theoretical investment in a Cartesian framework of the sort that typically seeks to establish the autonomy of the subject by defining it epistemologically as prior to any particular form of engagement with the world.

In Foucault's case, a similar logic is implicit in the assumption that discourses should be understood first and foremost as promoting epistemic frameworks of one kind or another, where these frameworks simultaneously influence our perception of reality and lead us to act, individually or collectively, in ways that make the world bear out the truth of our perceptions. The point here, as with Derrida, is that the conviction that this is *always* a problematic characteristic, though seemingly supported by historical studies of how particular structures of knowledge have been inflected by power relations, in fact requires something more than this, if it is to be credible as a *general* claim about human knowledge and the discourses that articulate it. What it requires is a basis for thinking that *any* self-validating epistemic framework is inherently repressive, and this thought also only seems intuitively compelling against the background of the Cartesian model, because it allows us to think that there is a sense in which it is meaningful to think of the individual subject as possessing some radically autonomous potential prior to any actual epistemic or non-epistemic relation to the world.

Such Cartesian thoughts lend credibility to a form of radical individualism that makes it easy to feel that a critical attitude towards norms of thought is inherently superior to an uncritical one. Yet this comes at a price. Wittgenstein's alternative account of language and meaning, by focussing on the conditions in which it becomes possible to distinguish meaningful signs from arbitrary accidental phenomena, alerts us to the fact that many of the ways in which we exist in relation to the world and to others, because they depend on language and concepts, also depend on the public contexts that make language and concepts comprehensible in the first case. For these forms of language and concepts to fulfil *that* role in our lives, they must, in many cases, correspond to a shared understanding, which for Wittgenstein means also a shared context of understanding. Such contexts are provided by communal practices and forms of life, but these need to be sustained through active participation – something that highlights a potential form of cultural disempowerment whose significance cannot be done justice to from the poststructuralist perspective. Typically, we tend to conceive of cultural disempowerment in terms of the idea that potentially autonomous individuals may be denied a voice, and so fail to get their particular identities, needs and values communally acknowledged, and poststructuralism seems – at first glance, at least – to offer an ideal theoretical framework for discussing this. But in our contemporary globalised environment, it may be that the more fundamental danger to our lives comes from the ways in which our dependence on communities for a basic sense of emotional connectedness to things – to the world and to other people – is increasingly frustrated and attenuated, as different forms of social breakdown make the active participation that would sustain these communities appear redundant to the individual members of those same communities.

My point, then, is that unconditional advocacy of the de facto superiority of critical to non-critical perspectives, as an alternative to conceding the outright incommensurability of cultural relativism and metaphysics, pushes us further in the direction of an equally unconditional embrace of the poststructuralist views on meaning and knowledge that can make this appear reasonable, and this also means an unconditional embrace of the radically individualistic Cartesian model of our relation to the world that legitimises poststructuralism itself. Yet this interposes an unconditional theoretical barrier between individuals and the conceptual resources they need to properly articulate a resistance to communal breakdown, since such resources must extend to thinking about the role that language and concepts play in our lives in terms that would have the potential to put this framework of unconditional commitments in question. This can be done, as Wittgenstein demonstrates, but only by refusing to incorporate Cartesian assumptions into any initial understanding of language and discourse. The fact that an unconditional preference for critical or relativist positions towards culture precludes such moves suggests that cultural relativism, in its own way, perpetuates the sort of covert discursive suppression of dissent that, for Foucault, was supported by all forms of rationality and metaphysics.

So adopting the self-reflexive hermeneutic approach as a basis for cultural relativism, and for rejecting "culture-transcendent perspectives" in a way that makes it look as though we have no alternative but to agree that all culture must be self-reproducing, puts cultural relativism itself in an impossible dilemma – one in which it is unable to fulfil its own critical social agenda. Either it must become so distanced from metaphysics that, as our consideration of the *Tractatus* has shown, it has no power to disrupt the latter's discourses, or it must become another version of metaphysics itself, implicated in the damaging social mechanisms that it has accused the latter of perpetuating.

Because the idea of culture as something not inherently self-reproducing may require us to see more than is visible from the perspective of *our* culture, which may indeed be self-reproducing for purely contingent historical and social reasons, it requires us to take up a position on whether "culture-transcendent perspectives" are possible. This looks at first sight like a straight choice between embracing traditional metaphysics and embracing the kind of cultural relativism that opposes traditional metaphysics as a matter of principle. But it should be clear by now that the idea that there is a meaningful choice to be made here is an illusion. Either the choice is arbitrary, as those two frameworks are simply incommensurable, or it is meaningless, as they both amount to versions of traditional metaphysics after all. So if we embark straightaway on discussions about *how* culture is self-reproducing, having decided that it *is* self-reproducing simply on the basis that any alternative stance marks a return to metaphysics (i.e. a potential endorsement of "culture-transcendent perspectives"), we are doing so on the basis of an arbitrary choice, and the idea that the self-reproducing nature of culture in general has been established as a real phenomenon that needs to be discussed is again an illusion.

Wittgenstein's work also suggests that we need not accept this arbitrariness. His account of the conditions under which it is possible to grasp that signs *have* a meaning suggests that even the most sceptical forms of critical discourse that seek to question *all* of our norms of thought presuppose a participation in shared linguistic practices, and thus in the communal forms of life that provide the necessary context for these. For somebody who finds his account persuasive, it must remain an open question how far these forms of life already commit one to a more or less conditional, or wholly unconditional, acceptance of such norms – or no acceptance of them at all. This thought allows us to turn our attention away from the dogmatic claims of both traditional metaphysics and cultural relativism, and to focus instead on actually *looking to see* which norms of thought we are committed to in practice, in *particular* areas or aspects of our life. That, I think, marks the first step on the road to a proper discussion of whether culture is, in fact, self-reproducing or not, since it suggests that such a discussion need not be constrained by the thought that it takes place within *our* culture, which, quite independently of this, may to some extent be distinguished by its self-reproducing tendencies.

Literary and Cultural Theory

General editor: Wojciech H. Kalaga

Vol. 1 Wojciech H. Kalaga: Nebulae of Discourse. Interpretation, Textuality, and the Subject. 1997.

Vol. 2 Wojciech H. Kalaga / Tadeusz Rachwał (eds.): Memory – Remembering – Forgetting. 1999.

Vol. 3 Piotr Fast: Ideology, Aesthetics, Literary History. Socialist Realism and its Others. 1999.

Vol. 4 Ewa Rewers: Language and Space: The Poststructuralist Turn in the Philosophy of Culture. 1999.

Vol. 5 Floyd Merrell: Tasking Textuality. 2000.

Vol. 6 Tadeusz Rachwał / Tadeusz Slawek (eds.): Organs, Organisms, Organisations. Organic Form in 19th-Century Discourse. 2000.

Vol. 7 Wojciech H. Kalaga / Tadeusz Rachwał: Signs of Culture: Simulacra and the Real. 2000.

Vol. 8 Tadeusz Rachwal: Labours of the Mind. Labour in the Culture of Production. 2001.

Vol. 9 Rita Wilson / Carlotta von Maltzan (eds.): Spaces and Crossings. Essays on Literature and Culture in Africa and Beyond. 2001.

Vol. 10 Leszek Drong: Masks and Icons. Subjectivity in Post-Nietzschean Autobiography. 2001.

Vol. 11 Wojciech H. Kalaga / Tadeusz Rachwał (eds.): Exile. Displacements and Misplacements. 2001.

Vol. 12 Marta Zajac: The Feminine of Difference. Gilles Deleuze, Hélène Cixous and Contemporary Critique of the Marquis de Sade. 2002.

Vol. 13 Zbigniew Bialas / Krzysztof Kowalczyk-Twarowski (eds.): Alchemization of the Mind. Literature and Dissociation. 2003.

Vol. 14 Tadeusz Slawek: Revelations of Gloucester. Charles Olsen, Fitz Hugh Lane, and Writing of the Place. 2003.

Vol. 15 Carlotta von Maltzan (ed.): Africa and Europe: En/Countering Myths. Essays on Literature and Cultural Politics. 2003.

Vol. 16 Marzena Kubisz: Strategies of Resistance. Body, Identity and Representation in Western Culture. 2003.

Vol. 17 Ewa Rychter: (Un)Saying the Other. Allegory and Irony in Emmanuel Levinas's Ethical Language. 2004.

Vol. 18 Ewa Borkowska: At the Threshold of Mystery: Poetic Encounters with Other(ness). 2005.

Vol. 19 Wojciech H. Kalaga / Tadeusz Rachwał (eds.): Feeding Culture: The Pleasures and Perils of Appetite. 2005.

Vol. 20 Wojciech H. Kalaga / Tadeusz Rachwał (eds.): Spoiling the Cannibals' Fun? Cannibalism and Cannibalisation in Culture and Elsewhere. 2005.

Vol. 21 Katarzyna Ancuta: Where Angels Fear to Hover. Between the Gothic Disease and the *Meat*aphysics of Horror. 2005.

Vol. 22 Piotr Wilczek: (Mis)translation and (Mis)interpretation: Polish Literature in the Context of Cross-Cultural Communication. 2005.

Vol. 23 Krzysztof Kowalczyk-Twarowski: *Glebae Adscripti.* Troping Place, Region and Nature in America. 2005.

Vol. 24 Zbigniew Białas: The Body Wall. Somatics of Travelling and Discursive Practices. 2006.

Vol. 25 Katarzyna Nowak: Melancholic Travelers. Autonomy, Hybridity and the Maternal. 2007.

Vol. 26 Leszek Drong: Disciplining the New Pragmatism. Theory, Rhetoric, and the Ends of Literary Study. 2007.

Vol. 27 Katarzyna Smyczyńska: The World According to Bridget Jones. Discourses of Identity in Chicklit Fictions. 2007.

Vol. 28 Wojciech H. Kalaga / Marzena Kubisz (eds.): Multicultural Dilemmas. Identity, Difference, Otherness. 2008.

Vol. 29 Maria Plochocki: Body, Letter, and Voice. Construction Knowledge in Detective Fiction. 2010.

Vol. 30 Rossitsa Terzieva-Artemis: Stories of the Unconscious: Sub-Versions in Freud, Lacan and Kristeva. 2009.

Vol. 31 Sonia Front: Transgressing Boundaries in Jeanette Winterson's Fiction. 2009.

Vol. 32 Wojciech Kalaga / Jacek Mydla / Katarzyna Ancuta (eds.): Political Correctness. Mouth Wide Shut? 2009.

Vol. 33 Paweł Marcinkiewicz: The Rhetoric of the City: Robinson Jeffers and A. R. Ammons. 2009.

Vol. 34 Wojciech Małecki: Embodying Pragmatism. Richard Shusterman's Philosophy and Literary Theory. 2010.

Vol. 35 Wojciech Kalaga / Marzena Kubisz (eds.): Cartographies of Culture. Memory, Space, Representation. 2010.

Vol. 36 Bożena Shallcross / Ryszard Nycz (eds.): The Effect of Pamplisest. Culture, Literature, History. 2011.

Vol. 37 Wojciech Kalaga / Marzena Kubisz / Jacek Mydla (eds.): A Culture of Recycling / Recycling Culture? 2011.

www.peterlang.de

Peter Lang · Internationaler Verlag der Wissenschaften

Wojciech Kalaga / Marzena Kubisz (eds.)

Cartographies of Culture

Memory, Space, Representation

Frankfurt am Main, Berlin, Bern, Bruxelles, New York, Oxford, Wien, 2010. 178 pp.
Literary and Cultural Theory. Edited by Wojciech Kalaga. Vol. 35
ISBN 978-3-631-60909-5 · hardback € 36,80*

Nowadays the issues of space and place pertain more than ever to the ongoing discussion about personal/regional/national identities. The worlds of private archives of memory often exist independently of political and administrative divisions, while dominant ideologies are often capable of re-defining national archives of memory through selective representation of the past. The way we remember our past and our heritage inscribes the space we live in: the places we remember and the places we wish to forget, the monuments we pull down and erect and the museums we build are only some of the signposts on the landscape of our cultural memory. The essays collected in this volume examine the role of places and spaces in the formation of both cultural practices and the existential experience of modern individual.

Contents: Urban Stories and Cityscapes · Spaces of Memory · Memory of Space · Margins, Borders and Ideologies · Spaces of the Visual

Frankfurt am Main · Berlin · Bern · Bruxelles · New York · Oxford · Wien
Distribution: Verlag Peter Lang AG
Moosstr. 1, CH-2542 Pieterlen
Telefax 0041 (0) 32 / 376 17 27

*The €-price includes German tax rate
Prices are subject to change without notice

Homepage http://www.peterlang.de